weather and climate

AF836809

25

weather and climate

J. HUME BROWN B. Sc.
Formerly Principal of Geography
now Assistant Rector
Clydebank High School

BLACKIE
London and Glasgow

Blackie & Son Ltd.,
Bishopbriggs Glasgow G64 2NZ
5 Fitzhardinge Street
London W1H ODL

© J. Hume Brown 1972

First published 1972
ISBN 0 216 89460 3

All rights reserved.
No part of this publication may be reproduced,
stored in a retrieval system, or transmitted,
in any form or by any means, electronic, mechanical,
photocopying, recording or otherwise, without
prior permission of Blackie & Son Ltd.

Printed in Great Britain by Robert Maclehose & Co. Ltd., Glasgow

Preface

The main purpose of this book is to examine the weather experienced in Britain (Part Two), and then to consider in rather less detail the principal types of climate occurring elsewhere in the world (Part Three). Before the weather and climate of specific places can be understood, it is necessary to have an elementary knowledge of the physical properties of the atmosphere and the factors causing variations in wind, temperature, rainfall, etc. These are considered in Part One.

Although the use of some technical terms is unavoidable, they are explained as simply as possible, and mathematical calculations are kept to a minimum. For those wishing fuller explanations there are several excellent books on meteorology and climatology to which reference should be made (see page 131).

Geography is basically the study of man and his environment, so reference to the influence of weather and climate on human activity is made wherever possible. Although detailed examination of particular places is beyond the scope of the book, it provides useful background information for the understanding of the climatic aspects of regional and sample studies.

Metric units are gradually coming into general use for climate statistics, but it will be many years before all atlases, textbooks and lists of statistics have been converted, so throughout this book both metric and imperial units are used. When converting from one set of units to the other, the figures have been rounded off to give numbers which are convenient to use, and therefore they are not always mathematically accurate conversions.

The text has been prepared to meet the requirements of G.C.E. 'A' level and Scottish Higher examinations, but it should also provide useful material for 'O' level candidates.

I wish to thank the following Examination Boards for permission to use questions from their past papers:
Oxford and Cambridge Schools Examination Board
Southern Universities Joint Board
Joint Matriculation Board
Scottish Certificate of Education Examination Board.

J.H.B.

Contents

Part THREE
World Climate

Introduction

Geography is the study of the various ways in which people live on the surface of the earth, and in particular of the influence of natural features (sometimes called the **environment**) on their lives. Three basic elements make up this natural environment: the land; the sea; and the air. It is the last of these which we are going to examine in this book.

The Atmosphere

The atmosphere is one of the most fundamental parts of our natural environment. Not only does it provide the life-giving air which is essential for the continued existence of creatures on the earth's surface, but it is also the source of our weather, which gives the varied conditions necessary for the complex assortment of plants and animals living in the world.

The atmosphere is mainly composed of gas, and it extends for many miles above the earth's surface. Its exact thickness is difficult to measure, as the density of the gas decreases rapidly with height, but traces of the atmosphere have been found as far as 1600 km (1000 miles) up.

STRUCTURE OF THE ATMOSPHERE

The atmosphere contains a number of layers of varying thickness which have fairly well-defined boundaries and different physical properties (fig. 1). Until

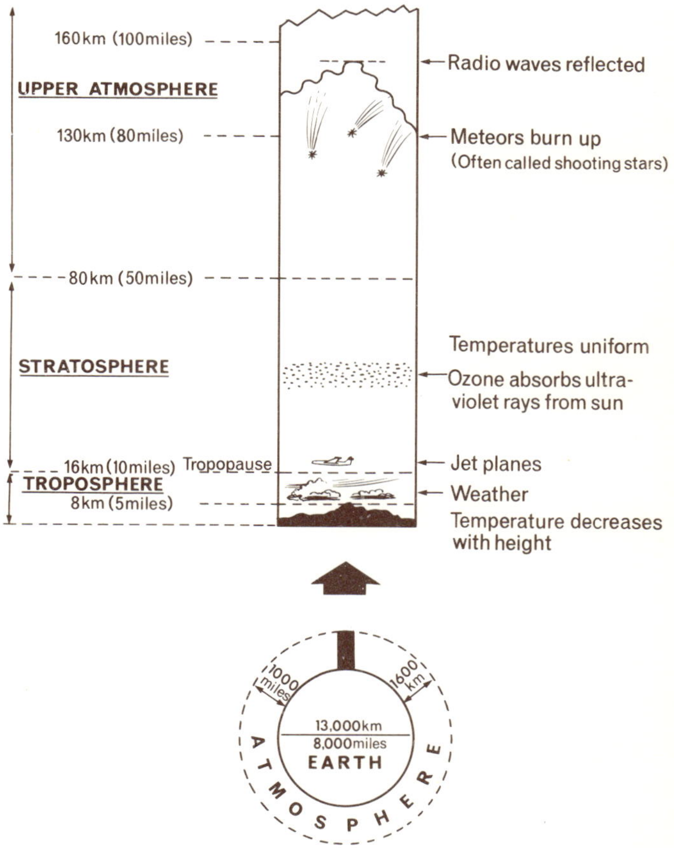

Fig. 1 The structure of the atmosphere

recently very little was known about the middle and upper layers, since information depended mainly on recordings made by instruments sent up on balloons, and these could only rise about 50 km (30 miles). Now instruments are sent to much greater heights on rockets and satellites, but our knowledge of the upper atmosphere is still very limited.

The lowest layer, called the **troposphere**, has a thickness of about 10 km (6 miles) at the poles, 11 km (7 miles) at latitude 50°, and 18 km (11 miles) at the equator. The troposphere contains about 85% of the gas and all the water in the atmosphere, and it is in this layer that our weather occurs.

The upper limit of the troposphere is called the **tropopause**, and above it lies the **stratosphere**, where temperature does not decrease with increase in height as it does in the troposphere. Above it lies the **upper atmosphere** whose density continues to decrease until finally the atmosphere ceases to exist.

Although these upper layers have no weather as we know it, they do have important effects on our lives: they absorb ultraviolet light from the sun; they burn up most meteors (small rock particles) entering the atmosphere from outer space; they are used in connection with radio communication; and we are now discovering that events in the stratosphere have a bearing on the weather of the troposphere.

Modern jet aircraft fly in the upper part of the troposphere, where winds are very different from those experienced near ground level. In this region of the troposphere they are mostly from a westerly direction (fig. 2).

In the stratosphere itself, the temperature no longer decreases with increasing altitude. The latitudinal pattern is the reverse of that experienced at the earth's surface: the air over the equator is colder than the air over the poles. This results in great contrasts in the air masses which meet at the tropopause and is the cause of turbulence in the atmosphere. This turbulence is particularly dangerous to aircraft, as without clouds there is no visible warning of its existence. Narrow belts of rapidly moving air called **jet streams** also occur at the tropopause. They are frequently associated with the depressions (Chapter 7) which

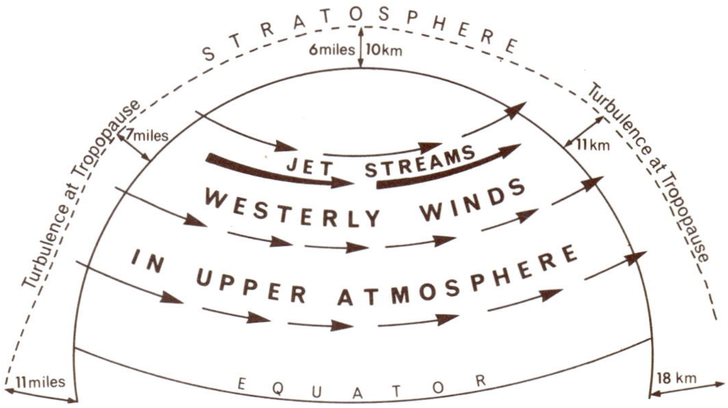

Fig. 2 Winds of the upper atmosphere

bring much of our weather, and they usually blow at speeds of between 240 and 480 kilometres per hour (150 and 300 m.p.h.) parallel to the fronts of the depression.

COMPOSITION OF THE ATMOSPHERE

The gas which we call *air* is in fact made up of a number of separate gases which occur in the proportions shown in the table:

Gas	*Volume percentage*
Nitrogen	78·09
Oxygen	20·95
Argon	0·93
Carbon dioxide	0·03
Other gases including neon, hydrogen and helium	0·003

These proportions are constant, but in addition the troposphere contains a gas called **water vapour** which varies in amount; some of it is always in the form of liquid water or ice, giving clouds, fog, frost, etc. Variations in the amount of water vapour cause much of the weather in the lower atmosphere.

Meteorology is the name given to the science which studies the weather. It is the changes constantly taking place in the atmosphere, (in particular in the troposphere) which cause variations in the weather throughout the world.

Weather can be described as the state of the atmosphere at any place at a particular time. Before a pattern can be seen in the weather and the variations in it explained, the meteorologist has to record its main elements. Among the most important elements he records are:

1 Atmospheric pressure
2 Relative humidity (the amount of water vapour present in the air)
3 Precipitation (rain, snow, fog, dew and mist)
4 Cloud types, heights and amounts
5 Visibility
6 Temperature
7 Sunshine
8 Wind speed and direction

These are all recorded in the lower atmosphere. As new techniques are developed, observations are being made of the upper parts of the atmosphere, where changes in temperature, pressure, and wind have important effects on the weather experienced on the earth's surface.

Climate When a record of the weather has been kept for a number of years (preferably more than 20 years), and averages of the various elements calculated, an annual pattern appears: this pattern is called the *climate*. The pattern differs in different parts of the world and, as a result, a number of different climates occur.

In the following sections we are going to look at:

Part One	The most important elements or factors which make up the weather
Part Two	The climate of the British Isles
Part Three	The main climates found throughout the world.

STUDY SECTION

1 What is another name for the air?
2 What is the *troposphere* and why is it so important to man?
3 State three ways in which the *stratosphere* differs from the troposphere.
4 State three ways in which the weather experienced by high-flying jet aircraft differs from that found at ground level.
5 List the main elements of weather recorded by the meteorologist.
6 What is the difference between weather and climate?
7 Why is it necessary for a geographer to have some knowledge of the work done by a meteorologist?

Atmospheric Pressure

1

Pressure and Altitude

The envelope of air which encloses the earth is pressing down on the earth's surface. The force with which it presses on unit area is called **atmospheric pressure**. This pressure is greatest at sea-level, where the whole thickness of the atmosphere is exerting its weight. The higher we go above the earth, the thinner becomes the layer of air pressing down, and therefore pressure decreases (fig. 3).

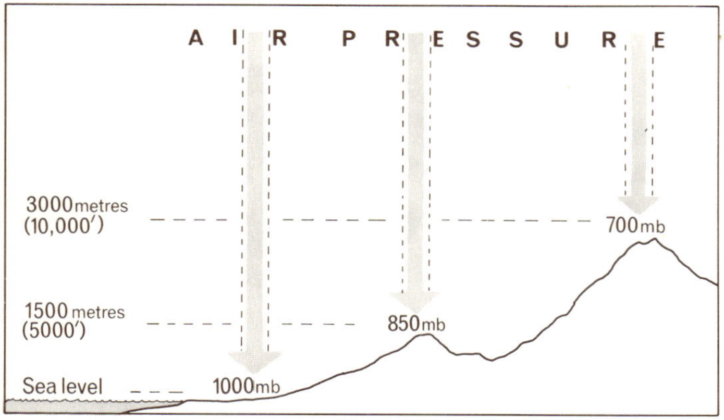

Fig. 3 How pressure decreases with height

At sea-level average air pressure is about 1 million dynes per square centimetre. This pressure is called 1 bar or 1000 **millibars** (mb).* 1 millibar represents a pressure of 1000 dynes per square centimetre, and this pressure decreases with height. In the lower atmosphere the decrease is approximately 10 mb per 100 metres. At altitudes of several thousand metres this figure becomes smaller, and the calculation is complicated as temperature changes affect the pressure.

A person on the earth's surface is surrounded by air which is pressing on him from all directions (fig. 4). His body is made to withstand it, and he only becomes conscious of air pressure when it changes markedly, for example, when he flies to a great height in an aircraft without any special protection. (In imperial units, air pressure measurements were

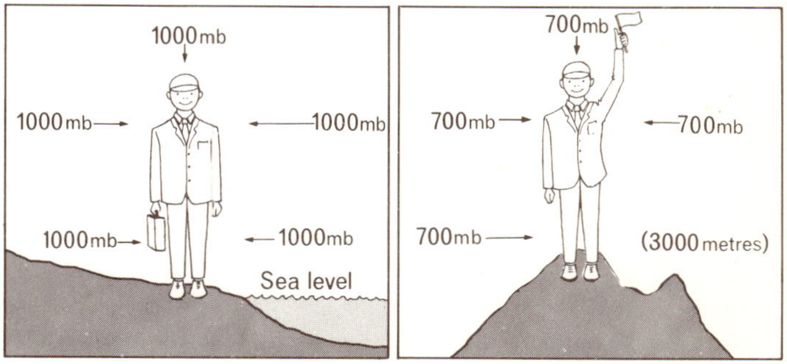

Fig. 4 Air pressure is the same in all directions

*In modern units 1 bar = 100,000 pascals (Pa) = 100,000 newtons per square metre (N/m^2). The scientist's standard pressure is 760 millimetres of mercury = 1013 millibars.

often given in inches, and the reason for this will be seen in the following paragraphs on the **barometer**.)

Measuring Air Pressure

Although air pressure at a given altitude is fairly constant, it does have slight but important variations which can only be measured by a very sensitive instrument called a **barometer**. There are two main types of barometer which are illustrated in figs. 5 and 6.

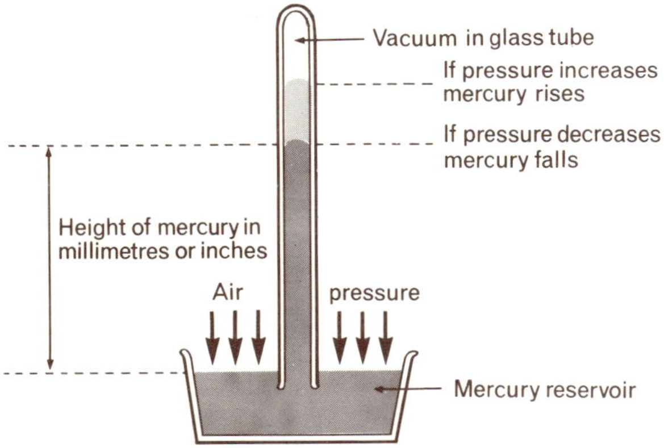

Fig. 5 Mercury barometer

In the **mercury barometer** the air pressing down on the surface of the mercury in the bowl forces some mercury to rise up the vacuum tube. The greater the air pressure, the higher the mercury rises up the tube. It is the height of this column in millimetres or inches which gives a measurement of air pressure. A column of mercury 750 millimetres or 29·5 inches long corresponds to a pressure of 1000 millibars.

In the **aneroid barometer** air is pressing on a metal box from which almost all the air has been removed. The sides of the box move in and out as air pressure increases and decreases, and these movements are transferred to a pointer and dial by a system of levers.

The mercury barometer is more accurate, but the aneroid barometer has these advantages: (*a*) it is much more portable and (*b*) the levers can be connected to a pen which traces a line on a revolving drum and thereby give a continuous record of pressure change.

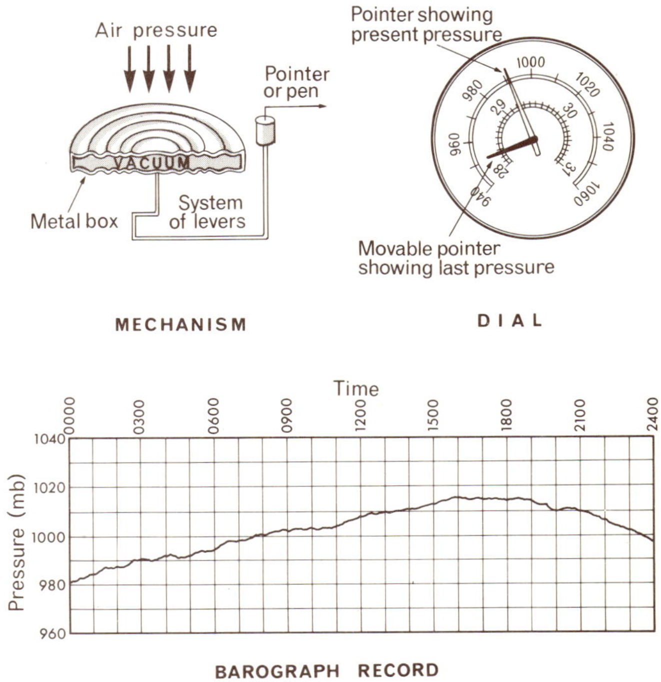

Fig. 6 Aneroid barometer

This constantly recording barometer is called a **barograph** and is valuable to the weather forecaster who wishes to know whether pressure is rising or falling.

An **altimeter** as used in aircraft or by some mountaineers is just an aneroid barometer which makes use of the decrease of pressure with altitude.

Recording Air Pressure

As with all weather measurements, the first record of air pressure consists of numbers noted in a table or on a graph; but before these numbers can be fully used by the meteorologist or the geographer, it is often desirable to transfer them to a map.

The stations at which pressure is measured are at many different altitudes owing to the uneven nature of most of the earth's surface. Before a map is made, all readings are reduced to sea-level by adding 10 mb for every 100 metres that the station is above sea-level. The pressure is then recorded on the map beside each measuring station. In order to see the pattern of pressure, lines called **isobars** are drawn through places of equal pressure. These lines are normally drawn at 2-mb or 4-mb intervals and, as can be seen in fig. 7, a certain amount of interpolation is necessary when drawing them.

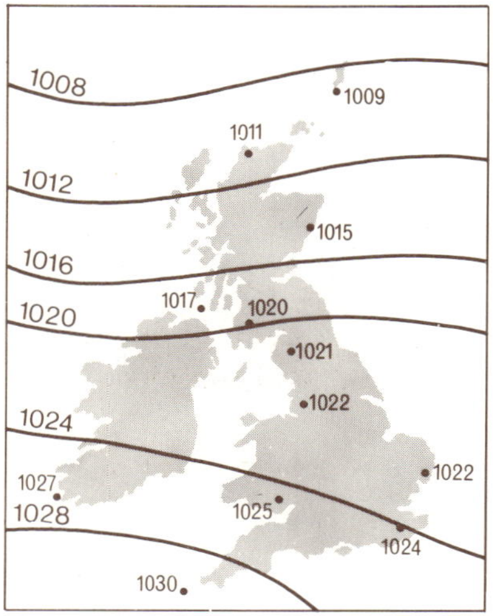

Fig. 7 How isobars are drawn by interpolation
Note that only a few of the recording stations are shown.

High and Low Pressure

In the troposphere, small changes in air pressure are always taking place. Although they are so small that we are not aware of them, they do have an important influence on the weather. These changes result in areas of **low pressure** (seldom lower than 950 mb) which are called **depressions** or **cyclones**, and areas of high pressure (seldom higher than 1050 mb) which are usually called **anticyclones**. The arrangement of the isobars in a typical depression and anticyclone are shown in figs. 8 and 9.

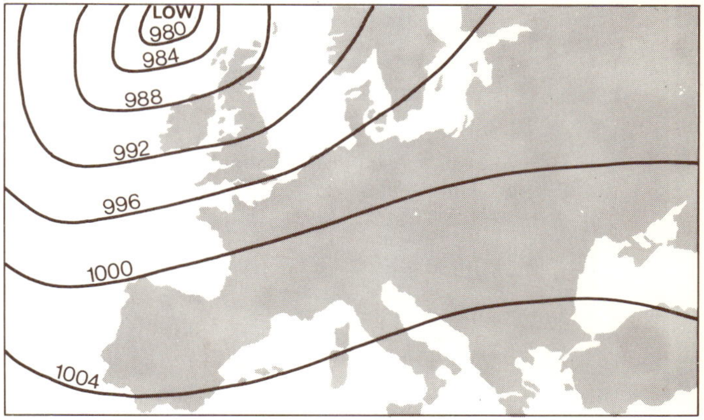

Fig. 8 Isobars for a depression centred off north-west Scotland

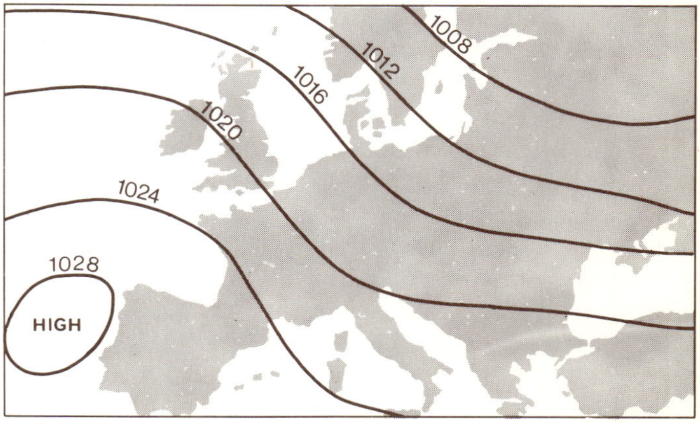

Fig. 9 Isobars for an anticyclone centred near the Azores

World Pressure Distribution

When air is heated it expands, becomes less dense, and therefore rises, reducing pressure in the lower atmosphere. As we shall see in a later chapter on temperature, the air receives most of its heat from the earth's surface. Since this heat is unevenly distributed over the earth, the lower air is heated unevenly. This uneven heating results in uneven distribution of air pressure and, although the basic pattern is simple, with low pressure at the hot equator and high pressure at the cold poles, it is complicated in the following ways:

1 There is a second high-pressure area in the mid-latitudes (30°—40°) of each hemisphere (fig. 10).

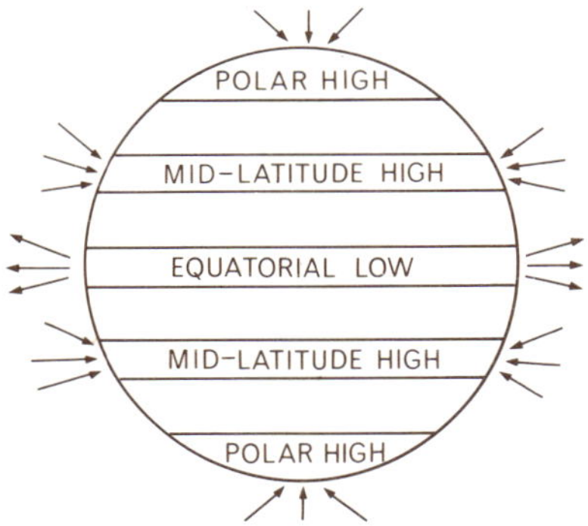

Fig. 10 Main areas of ascending and descending air and the associated pressure belts

2 These mid-latitude high-pressure areas are broken up in summer by low pressure over the hot continents. This feature is more marked in the northern hemisphere, where there are larger land areas, than in the southern hemisphere.

3 The seasonal migration of the sun north and south of the equator causes the area of equatorial heat, and therefore the equatorial low-pressure belt, to migrate with the seasons.

These pressure distributions are illustrated in figs. 11 and 12.

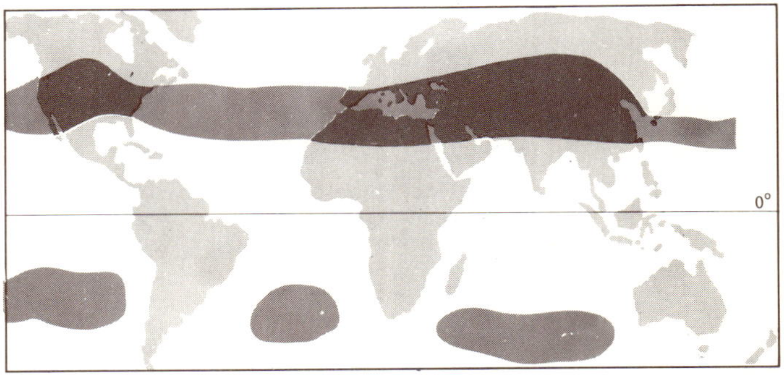

Fig. 11 Main high-pressure areas in January
Note how continents in the southern hemisphere break up mid-latitude belt when they heat up in summer

Fig. 12 Main high-pressure areas in July
Note how northern hemisphere continents break up mid-latitude belt when they heat up in summer

STUDY SECTION ON CHAPTER 1

1 Why is air pressure greater at sea-level than on high land?

2 Draw a diagram showing a mercury barometer and explain how it measures air pressure.

3 What is a barograph?

4 What is an isobar?

5 Give an alternative name for (*a*) a low-pressure area, (*b*) a high-pressure area.

6 Draw a diagram showing the main belts of high and low pressure in the world.

7 Why do these pressure belts move north in June and south in December?

8 Is central Asia an area of high pressure or low pressure in December?

9 Trace the accompanying map which shows the air pressure recorded at a number of places in and around Britain. On your tracing:

(*a*) Draw isobars at 4-mb intervals from 976 mb to 1016 mb.

(*b*) Beside each isobar give its pressure in millibars.

(*c*) Print HIGH PRESSURE and LOW PRESSURE once each at appropriate places.

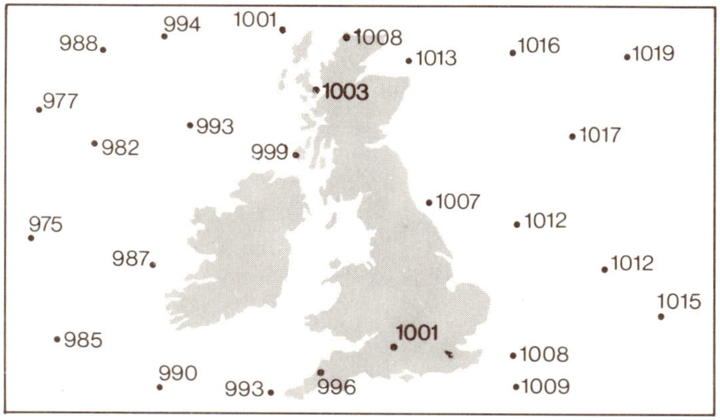

Wind

2

In the previous chapter we have seen that air pressure varies from place to place and from season to season. This uneven distribution causes air to move from one part of the atmosphere to another. Some of these movements are of a vertical nature and we call them **air currents**; others are predominantly horizontal, and we call them **winds**. Winds blow throughout the troposphere, and often the air higher up is blowing in a different direction from air near the ground. Although these upper winds have a bearing on our weather and are of great importance to airmen, it is the wind near the ground which has the most direct influence on our weather and which in fact is part of it.

Speed and Direction

The direction given to a wind is that from which it is blowing. It is observed on a **wind vane** (fig. 13)

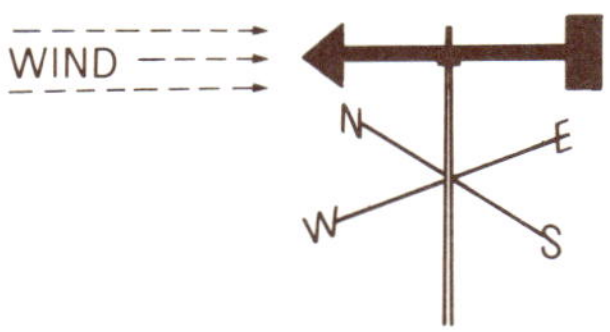

Fig. 13 Diagram of a wind vane

where a finely pivoted arrow swings to point in the direction from which the wind is coming. The vane should be situated where it can show the true wind direction and is not affected by eddies coming round buildings or from high roofs. The direction is given in degrees 0°—359° or in compass points.

Wind speed is measured by an **anemometer** (fig. 14). This instrument consists of cups on the ends of

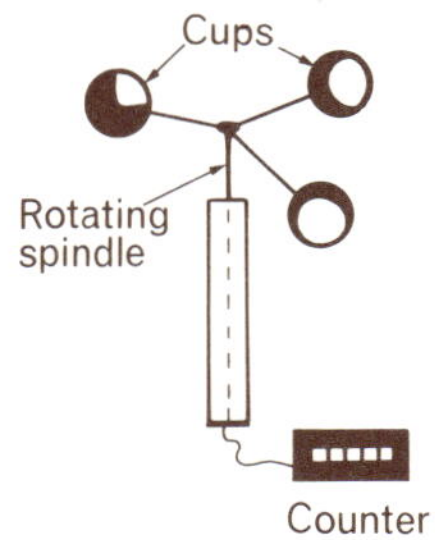

Fig. 14 Diagram of an anemometer

arms which are connected to a spindle. When the wind blows into the cups the spindle rotates, and the speed of rotation measures the speed of the wind. This measurement is given in knots, kilometres per hour, miles per hour, or metres per second.

Wind speed can also be estimated by using the **Beaufort scale**. It was originally devised for use by seamen in the days of sailing ships, but has now been adapted for land use also.

BEAUFORT SCALE OF WIND FORCE

Force number	*Description*	*Effect*	*Speed k.p.h.*	*Speed m.p.h.*
0	Calm	Smoke rises vertically	less than 1	
1	Light air	Air movement shown by smoke but not by wind vanes	1–5	1–3
2	Light breeze	Wind felt on face; leaves and wind vane move	6–11	4–7
3	Gentle breeze	Small twigs move and light flags flutter	12–19	8–12
4	Moderate breeze	Raises dust and paper; small branches move; flags fly	20–29	13–18
5	Fresh breeze	Small trees in leaf sway	30–39	19–24
6	Strong breeze	Wind whistles in telegraph wires; large branches move; umbrellas difficult to control	40–50	25–31
7	Moderate gale (near gale)	Whole trees move; walking hampered	51–61	32–38
8	(Fresh) gale	Twigs break off trees; walking made difficult	62–74	39–46
9	Strong (severe) gale	Slates blown off roofs	75–87	47–54
10	Whole gale (storm)	Trees uprooted; overhead wires blown down; considerable structural damage	88–101	55–63
11	Storm (violent storm)	Widespread damage to buildings, communications, and woodlands	102–120	64–75
12	Hurricane	Devastation occurs	over 120	over 75

Pressure and Wind

In a motor-car tyre the air is at a higher pressure than the surrounding air. If the tyre is punctured (fig. 15) air moves out from the high-pressure inside; in a similar way air moves out from high-pressure areas of

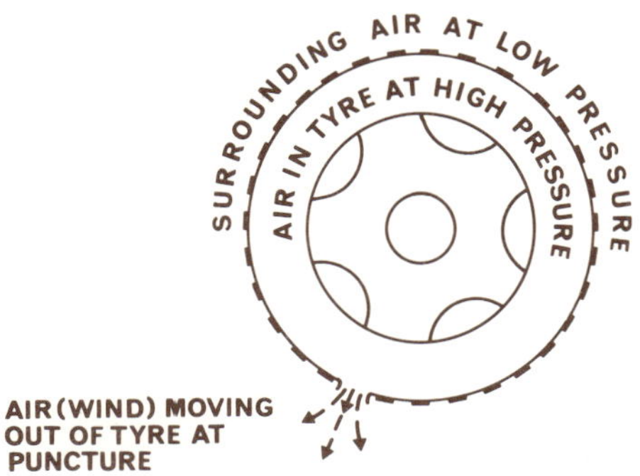

Fig. 15 Air moving from high pressure to low pressure

the atmosphere. Because the earth is rotating (with its atmosphere) however, this air movement (wind) is deflected (fig. 16). In the northern hemisphere the wind is deflected to the right, and in the southern hemisphere to the left. As a result, winds tend to blow round pressure centres, that is, roughly parallel to the

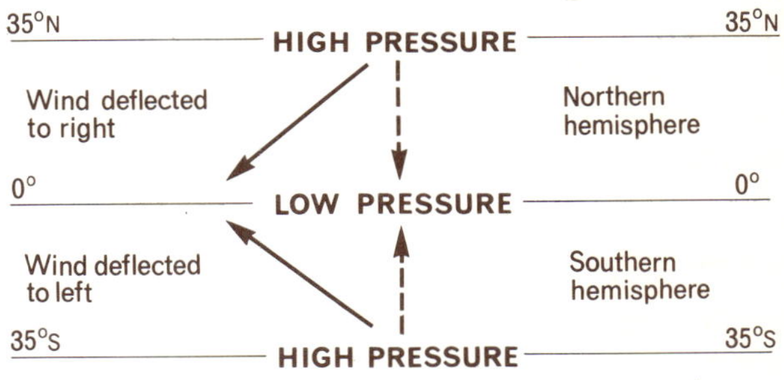

Fig. 16 How wind is deflected by the rotation of the earth

isobars (fig. 17). The connection between wind direction and pressure is given in **Buys Ballot's Law**. It states that, if you stand with your back to the wind, Low pressure is to your Left in the northern hemisphere (and to your right in the southern hemisphere).

As well as causing winds to blow and influencing their direction, the pressure distribution affects the

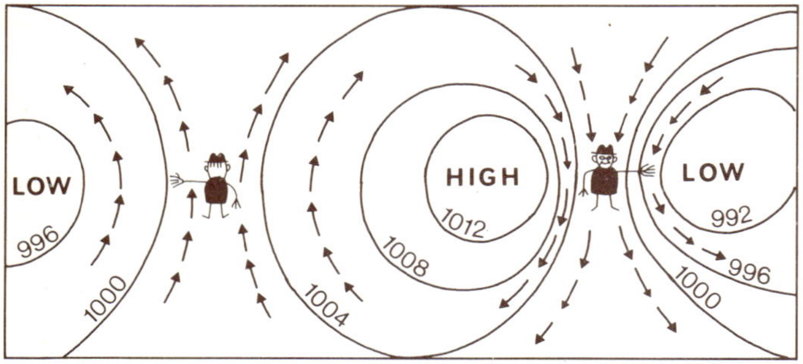

Fig. 17 If you stand with your back to the wind, low pressure is to your left in the northern hemisphere (Buys Ballot's law)

wind speed. When pressure changes greatly over a short distance, there is said to be a steep **pressure gradient**. This is shown on a map by isobars being close together, and strong winds occur. On the other hand, a gentle gradient where the pressure change is spread over a much longer distance results in light winds. The connection between pressure gradient and the spacing of isobars (fig. 18) is very similar to that between steep and gentle slopes on the land and the spacing of contour lines on an Ordnance Survey map.

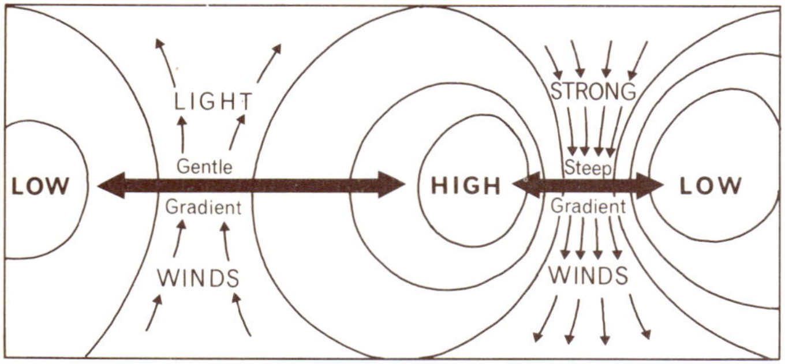

Fig. 18 Connection between pressure gradient and wind strength

Change of Direction

As pressure systems move about the atmosphere, they bring changes of wind direction to the lands over

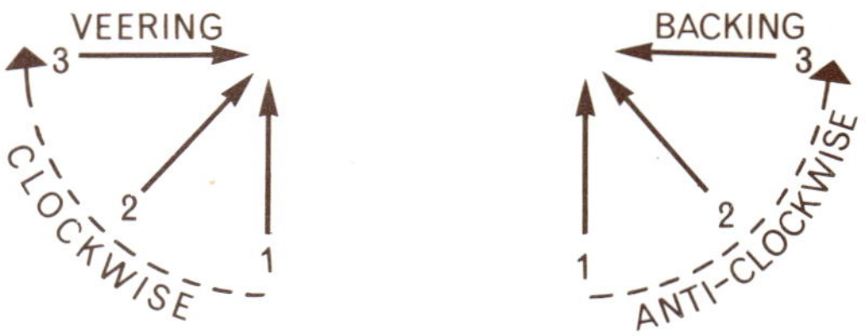

Fig. 19 Veering winds change in a clockwise direction, and backing winds change in an anticlockwise direction

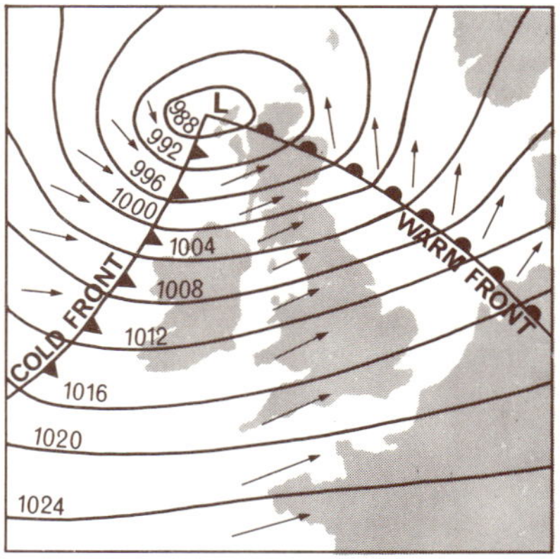

Fig. 20 Depression and its winds at 12.00 hours

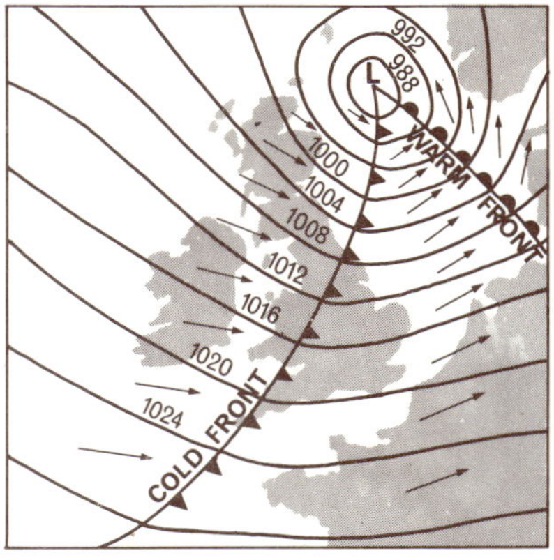

Fig. 21 The same depression and its winds at 18.00 hours

which they pass. If this change is clockwise, the wind is said to be **veering**; if the change is anticlockwise, the wind is said to be **backing** (fig. 19). Figs. 20 and 21 show how, as a depression crosses Scotland, winds change direction from west to north-west with the passage of the cold front.

Wind Frequency

In all parts of the world, some winds are more common than others; these are known as **prevailing winds**. In many tropical areas, on the other hand, the wind blows from the same direction almost all year, and these winds are called **trade winds**. In **monsoon** areas there are usually two prevailing winds which come from opposite directions; each blows for about half the year.

The direction of the prevailing wind at sea was very important in the days of sailing ships, as most of them found great difficulty in sailing into the wind. This is most clearly seen in the triangular route (fig. 22) which

Fig. 22 Route followed by sailing ships in the North Atlantic

was used by ships trading across the North Atlantic. For the westward voyage to America they kept well to the south, in order to use the easterly trade winds, while on the return voyage to Europe they kept further north in the region of the westerlies.

The Arabs of East Africa use the reversal of the monsoon winds (fig. 29) when trading with India by sailing there during the SW monsoon and returning to East Africa when the NE monsoon blows. In more recent times Sir Francis Chichester and the other 'round-the-world' single-handed yachtsmen made use of various prevailing winds during their voyages as is shown on fig. 23. Chay Blyth on the other hand

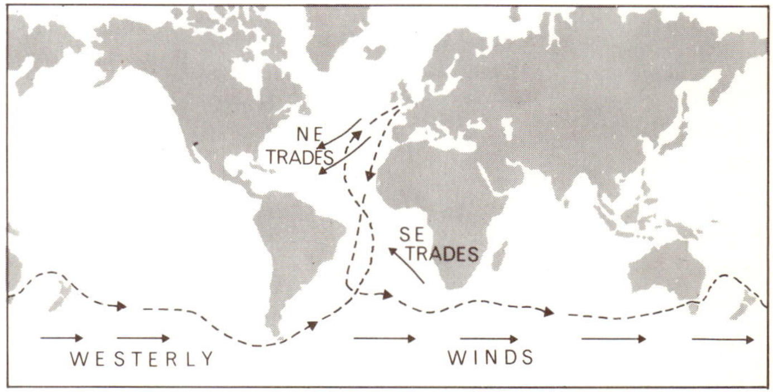

Fig. 23 Route followed by Sir Francis Chichester on his voyage round the world.

N	NE	E	SE	S	SW	W	NW	Calm
3·9	12·2	9·1	5·6	10·3	19·4	13·6	7·8	18·1

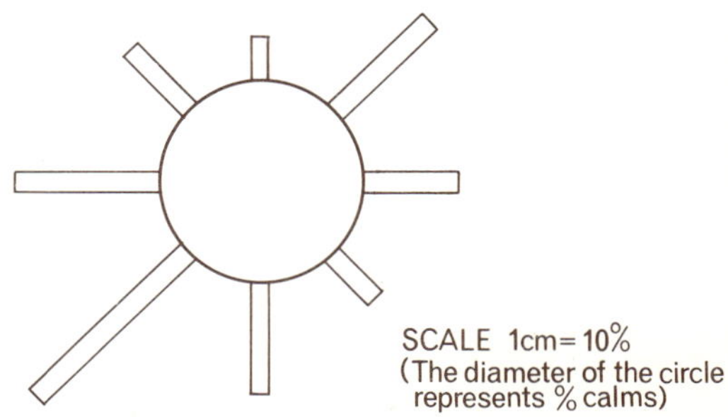

Fig. 24 Wind rose for Glasgow
Percentage frequency of wind direction

chose to fight the prevailing winds by sailing against the Roaring Forties. These produce some of the stormiest seas and fiercest gales in the world, since there is hardly any land to break their force.

The wind frequency for a place can be illustrated by a **wind rose** (fig. 24). It is a diagram where lines are drawn from a central point along each of the main compass directions. The length of each line is proportional to the frequency of the wind from each direction.

Major World Winds

The pattern of world pressure distribution described in Chapter 1 and illustrated in fig. 10 results in a pattern of air currents and major winds which are shown in figs. 25 and 26. In fig. 25 it will be seen that,

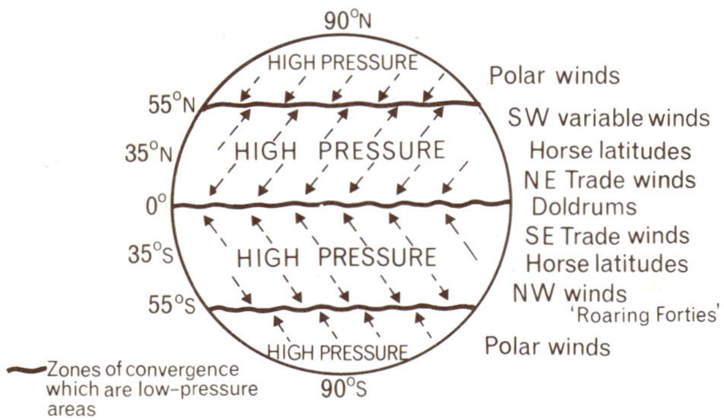

Fig. 25 Major world wind belts

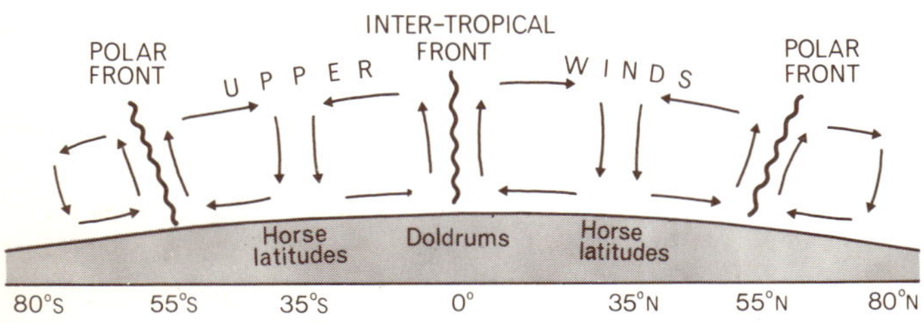

Fig. 26 Connection between surface winds, vertical currents, and upper winds

although the winds tend to blow from high to low pressure, they are deflected to the right in the northern hemisphere and to the left in the southern hemisphere. Fig. 26 shows that high-pressure areas have descending air and low-pressure areas have ascending air. These vertical movements have an important influence on the weather, since rising air tends to produce unsettled wet weather while descending air tends to give sunny dry conditions.

The three areas where the air starts to rise are called **zones of convergence**, as two different air masses meet at them; the irregular boundary between these masses is called a **front**. In the case of the **polar fronts**, the two air masses which meet differ greatly in their temperature, humidity, etc., since one originates near the tropics while the other comes from the polar ice cap. Rotating storms called **cyclones** are formed in the latitudes of these fronts as a result of the meeting of these two contrasting bodies of air. (We have seen in Chapter 1 that a cyclone is also an area of low pressure.)

Similar cyclones occur at the **intertropical front** but, since both air masses are of tropical origin, they are more alike and the storms are normally less severe. This relatively straightforward pattern of winds is complicated in two important respects:

1 The annual north-south migration of the pressure belts with the overhead sun (fig. 27) results in a corresponding migration of the wind belts (fig. 28). Many places on the earth's surface have one pressure and wind regime in summer and a different one in winter. This north-south migration is sometimes called the **swing of the wind belts**, and is an important factor in causing the dry summers and wet winters of areas with a Mediterranean type of climate (see Chapter 13).

2 The uneven distribution of land and sea, which results in the partial collapse of some of the high-pressure belts in summer (as shown on figs. 11 and 12), causes seasonal continental winds

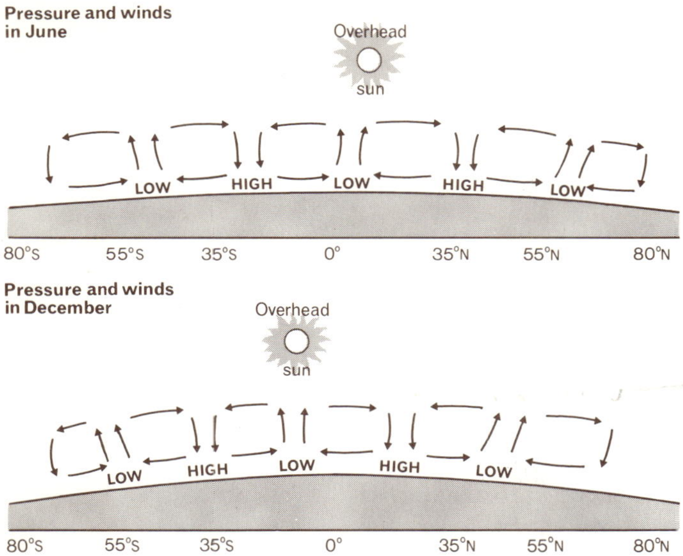

Fig. 27 Migration of pressure belts and winds with the overhead sun

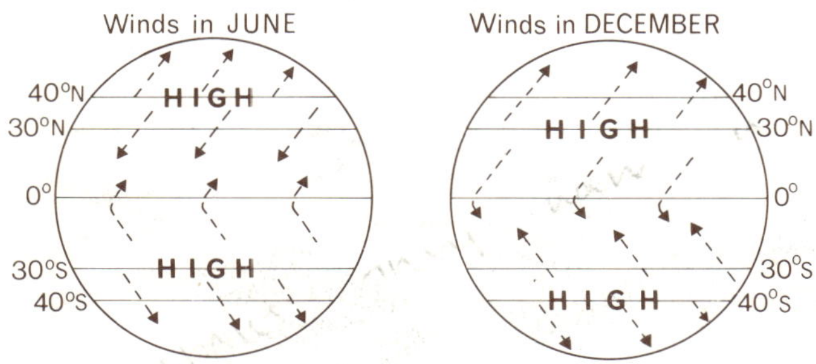

Fig. 28 The swing of the wind belts
Note 1. The area between 30° and 40° latitude changes its wind system according to the season
2. When winds cross the equator they change direction by 90°

which do not follow the main world pattern. This is particularly marked in the northern hemisphere where the proportion of land is relatively great. The low pressure of the interior of North America and Eurasia allows damp marine air to penetrate far inland, giving summer rain.

Regional and Local Winds

In certain parts of the world special local conditions lead to the development of regional winds which affect fairly large areas. Sometimes they have a marked seasonal duration.

1 MONSOON WINDS

Monsoon winds occur in the tropics, where a large land mass lies on the polar side of an equatorial ocean. These conditions occur in West Africa, North Australia, and especially in the Indian sub-continent. The Indian monsoon winds are illustrated in fig. 29. During summer, great heat in northern India results in very low pressure and strong on-shore winds blow.

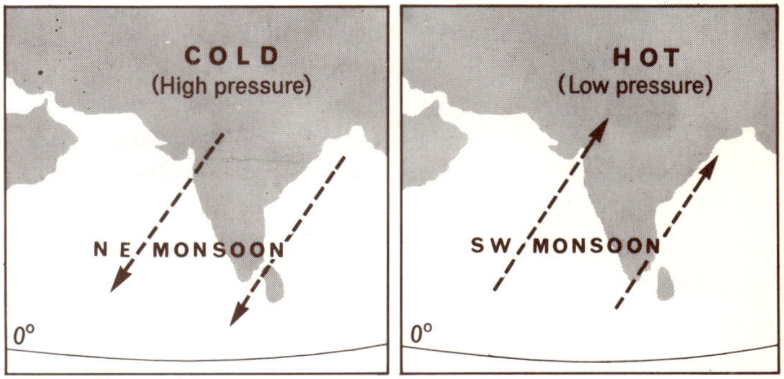

Fig. 29 The 'Indian' monsoons
left conditions during winter
right conditions during summer

In winter the winds reverse direction (this is characteristic of monsoons), since the land is much cooler and becomes an area of high pressure. The characteristic deflection to the right can be seen in both summer and winter monsoons. The south-west monsoon is the main source of water for the Indian farmer, and its failure can result in famine.

2 HARMATTAN

The **harmattan** is a hot dry wind which blows from the Sahara desert to the West African countries between November and March. Although it often brings clouds of red dust, it is very welcome after the hot damp conditions of the rest of the year, and is sometimes locally known as *the doctor*.

3 SIROCCO

The **sirocco** is another hot dry dusty wind which blows from the Sahara, but this time it affects the Mediterranean coasts of both Africa and Europe (fig. 30). It normally blows in front of a depression moving east along the Mediterranean. It brings unpleasant hot dusty conditions to Algeria, where it can cause considerable damage to crops. By the time it reaches the shores of Europe it has picked up considerable moisture and its high humidity is very depressing. The same wind is called the **leveche** in Spain and **khamsin** in Egypt.

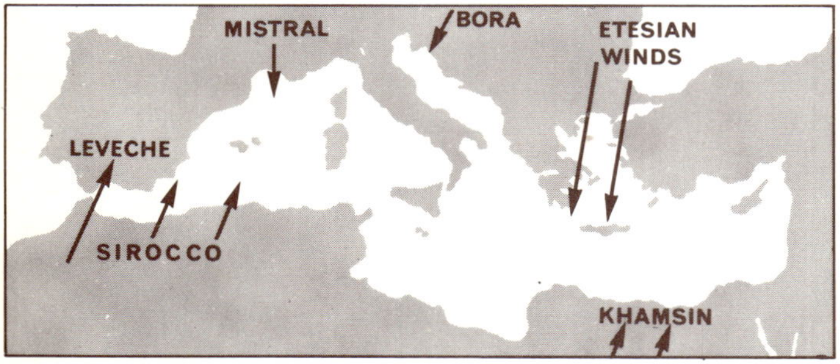

Fig. 30 Some local winds of the Mediterranean region

4 MISTRAL

Northerly winds which blow at the rear of a Mediterranean depression are sometimes strengthened by high pressure over continental Europe. This results in a bitterly cold dry wind called the **mistral** funnelling down the Rhone valley (fig. 30). It brings unseasonably low temperatures which damage crops. Hedges of cypress are often planted along the north sides of fields to protect orchards.

5 BORA

The **bora** is a similar northerly wind at the head of the Adriatic (fig. 30).

6 ETESIAN WINDS

Etesian winds are northerly gales which blow in the eastern Mediterranean in summer (fig. 30). They make sailing dangerous and prohibit the growing of trees in exposed situations.

7 FÖHN

The **föhn** is a hot dry wind which blows from the Alps down valleys in Switzerland and Austria (fig. 31).

Fig. 31 Föhn wind

When it occurs, the temperature may rise as much as 15°C in 24 hours. Because of its strength and drying effect, special restrictions regarding the lighting of fires in the open are often enforced. It causes rapid melting of snow.

8 CHINOOK

The **chinook** which blows from the Rocky Mountains to the Canadian Prairies is another example of a föhn wind. It raises the winter and early spring temperatures of the High Plains above freezing, melts the snow cover, and makes valuable pasture available for grazing much earlier in the year than it would be otherwise.

9 CYCLONIC WINDS

Cyclonic winds are wind systems which blow round low-pressure centres. Their strength depends largely on the intensity of the depression (or low-pressure centre) and the pressure gradient which results. In the regions of the polar and intertropical fronts these winds occur in depressions. In certain other parts of the world, steeper pressure gradients give rise to rotating storms called **hurricanes** (fig. 32a) (in the

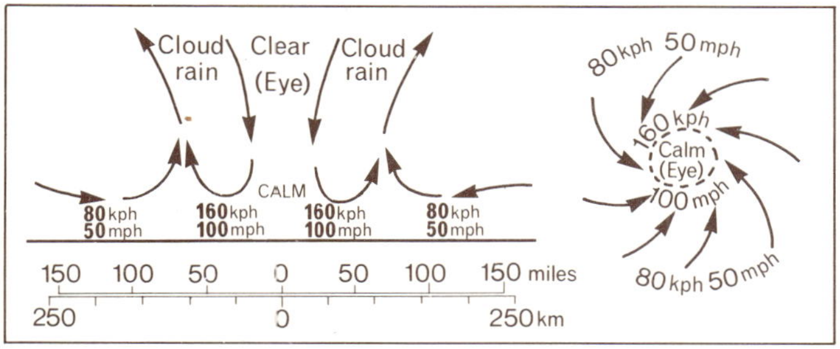

Fig. 32a Cross-section and plan of a hurricane

West Indies), **typhoons** (in the Philippines), and **cyclones** (off the south-east coasts of Africa and Australia). In these storms the centre or **eye** of the low-pressure systems is often an area of calms or light winds. Winds of the hurricane type are most destructive and cause severe structural damage on land, as well as being a danger to shipping.

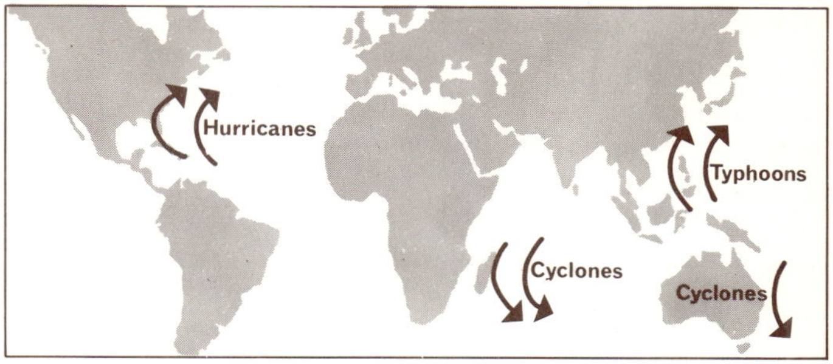

Fig. 32b Regions where tropical storms are common

10 TORNADOES

Tornadoes are very small areas of extremely low pressure with winds swirling very rapidly round the centre. They are often only about half a mile in diameter. As they move along, they carve a path of destruction, often lifting complete buildings off the ground. They are common on hot plains such as those of the central United States in summer. Over the sea they cause waterspouts.

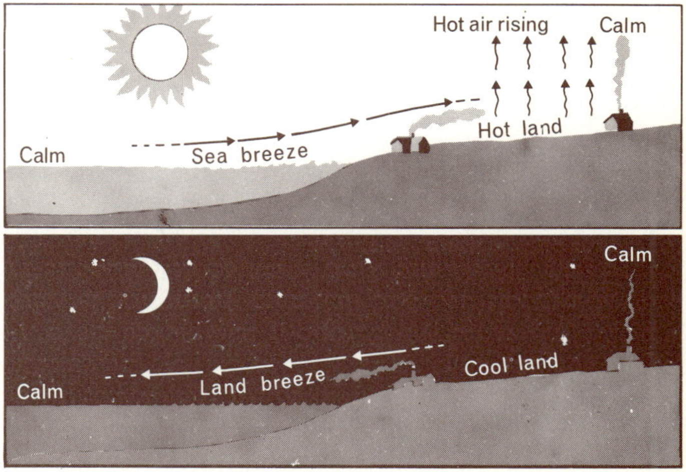

Fig. 33 Land and sea breezes

11 LAND AND SEA BREEZES

During a summer anticyclone, when the pressure gradient is very gentle and there is no general wind, local coastal breezes can develop (fig. 33).

During the day the land heats more rapidly than the sea, and air rises from the land. Air moves in from the sea to replace this rising air and a **sea breeze** develops.

At night the land cools more rapidly than the sea. The cool air over the land flows downhill to the sea giving a **land breeze**.

Both winds are fairly gentle and extend only a few few miles on either side of the coast.

STUDY SECTION ON CHAPTER 2

1 In which direction would you be facing if you had your back to a north-west wind?

2 What is the Beaufort wind scale?

3 Give the following wind speeds as *forces* on the Beaufort Scale, and describe how you would recognize them without the use of instruments:

10 m.p.h., 53 km.p.h., 50 m.p.h.

4 Why does wind not blow directly from high pressure to low pressure?

5 What does Buys Ballot's law tell you?

6 Trace the following diagram and add to it:
(*a*) arrows to show the wind direction at A, B, C;
(*b*) the words *light*, *moderate*, *strong*, where these words would describe the wind strength.

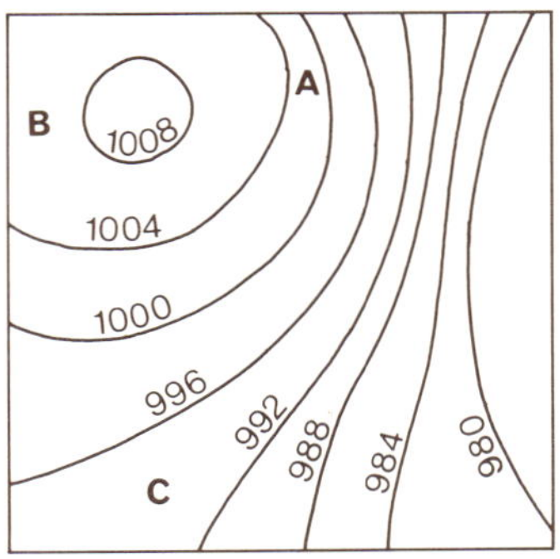

7 If wind direction changes from south to south-west, will it be veering or backing?

8 Study the wind rose for Glasgow (fig. 24).
(*a*) What is the prevailing wind for Glasgow?
(*b*) What is the least common wind for Glasgow?

9 Draw a wind rose from the following statistics:

MEAN WIND DIRECTION AT DURBAN IN JULY

N	3%	SE	4%	W	27%
NE	23%	S	4%	NW	3%
E	12%	SW	20%	CALM	4%

10 After studying figs. 25 and 28, state the prevailing wind direction at: 15°N; 50°S; 35°N in July; 35°S in July.

11 Explain fully the swing of the wind belts.

12 What are the **doldrums** and where would you find them?

13 Which of the following winds would you find at places lettered 1 to 6 on the accompanying map of the world? If the winds are seasonal, state one month when they would be blowing.

A—SW monsoon
B—sirocco
C—chinook
D—hurricane
E—harmattan
F—föhn

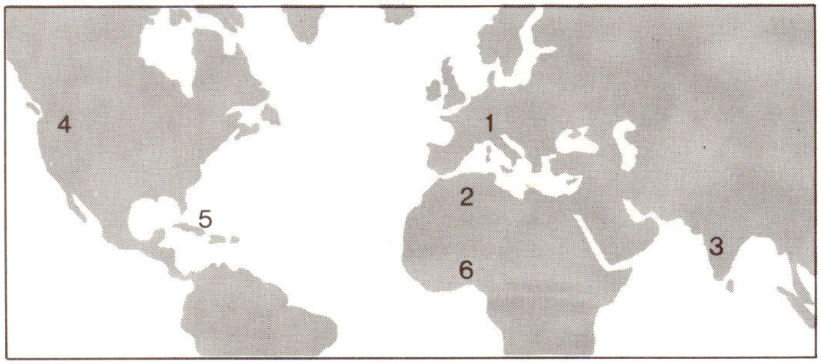

14 Make the following additions to the map you drew for question 9 in Chapter 1:
(*a*) Draw arrows to show wind directions.
(*b*) Print *strong wind*, *moderate wind*, and *light wind*, once each at appropriate places.

15 State what the *polar front* is, and explain why it is a region of disturbed weather.

16 Explain why a yacht close to the shore is likely to sail faster than a yacht a few miles off shore, during a summer anticyclone.

Ocean Currents

3

When winds blow over an ocean there is a certain amount of friction between the air and the water. As a result the surface layers of the sea start moving in the direction of the prevailing wind. In some cases this movement is slow but widespread, giving rise to an **oceanic drift**; in other cases the shape of the coast and the configuration of the sea bed result in a well-defined stream, flowing with a measurable speed through the ocean to form an **ocean current**.

These drifts and currents take water from their area of origin for thousands of miles. If they are flowing to or from the equator, they cause the sea temperature at their destination to be very different from that which would be expected for the latitude.

The general pattern of these currents and drifts shown on fig. 34 illustrates two important features:

1 In northern oceans the general circulation is clockwise, while in southern oceans it is anti-clockwise.

2 Currents flowing from the equator bring warm water, while currents flowing towards the equator bring cold water.

As well as affecting the temperature of the seas, these currents have an important influence on the temperature of the lands whose shores they wash, especially if the prevailing wind is on-shore. The influence of coastal currents is particularly well marked between the equator and about 40° latitude, where the east coasts of the continents are washed by warm currents, while the west coasts are affected by cold currents. Beyond 40°S there is relatively little land, but in the northern hemisphere, where there are large land masses north of 40°N, the currents continue to affect the climate. Here it is the west coasts which are kept mild, while the east coasts are affected by the cold currents.

When fig. 34 is compared with the wind belts shown on fig. 25 (page 14), it will be seen that the low-latitude cold currents occur in regions of off-shore winds—the trade winds. Examples are the West Australian current, the Benguela current off the west coast of South Africa, and the California current on the west coast of North America. These trade winds move surface water away from the coast, and it is replaced by an upwelling of water from the depths of the sea (fig. 35). This rising water is cold and contributes greatly to the low temperature of the currents.

The influence of a cold current on the lives of people living in its vicinity is particularly well seen in coastal Peru, where the north-flowing current is about 80 km (50 miles) wide and has an average temperature of 15°C. Its most obvious influence is seen in the average temperatures of the coastal towns. They are about 7°C lower than those experienced at the same latitude in coastal Brazil. This current also contributes to the drought of the region. Although the prevailing

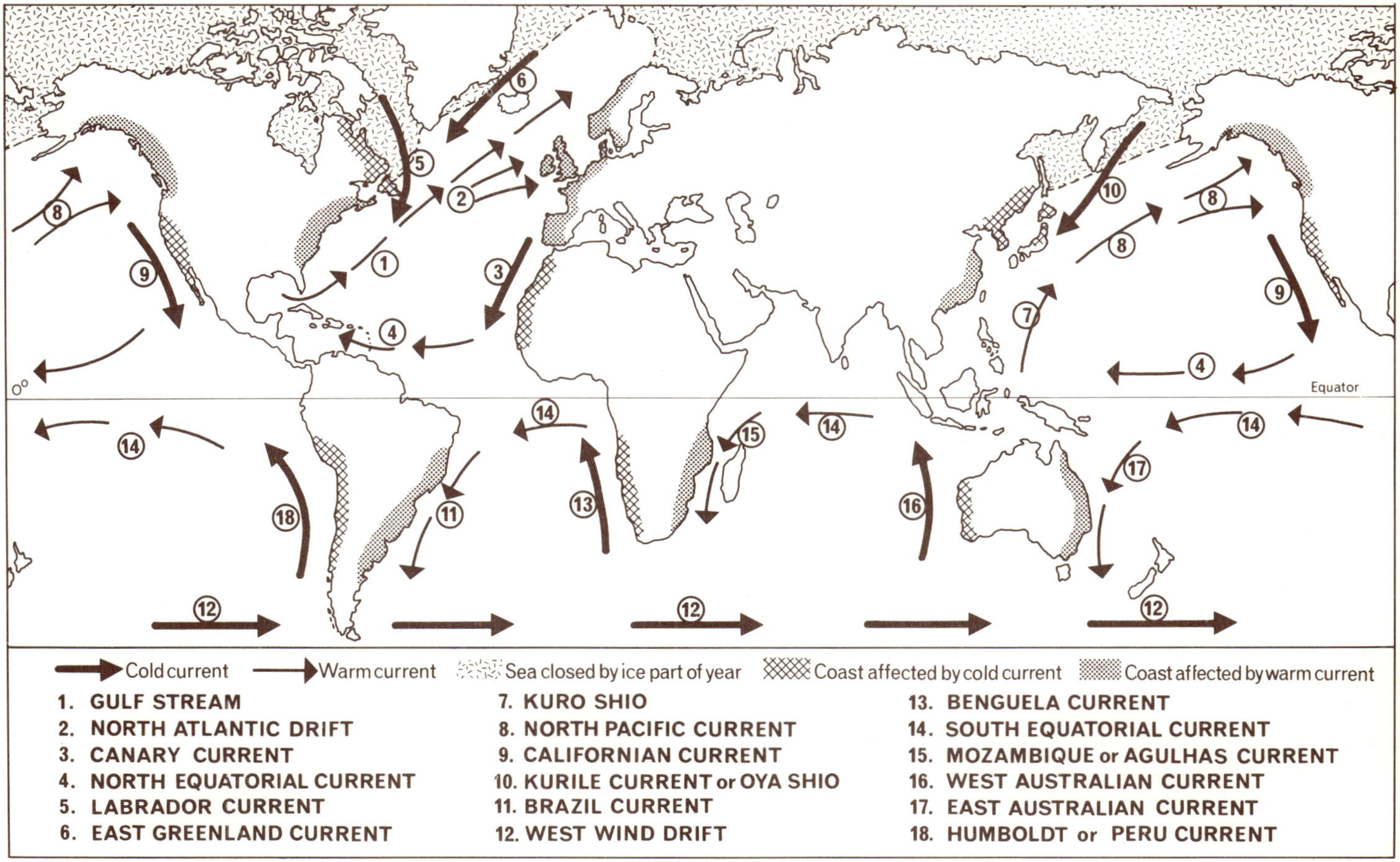

Fig. 34 Map of ocean currents and drifts

wind is off-shore, it does occasionally blow from the sea. When this happens the moist marine air loses some of its water vapour, in the form of coastal fog, on crossing the cold current; when it reaches the slightly warmer land its temperature rises to give very dry air. Lima has an average annual rainfall of only 50 mm (2 inches) and there are many years with no rain at all. The upwelling water from the depths of the ocean is chemically very rich. As a result, marine life is abundant, with vast shoals of fish which provide food for the people and for millions of sea birds. The guano collected from the nests of these birds is a valuable fertilizer which is used to enrich the soil of the irrigated coastal farms. It also provides a valuable export.

Occasionally the coastal current is deflected away from the coast by warmer water from the north. Without the cold current to cool the air and reduce its humidity with fog formation, the moist sea air reaches the land in a highly saturated state. This

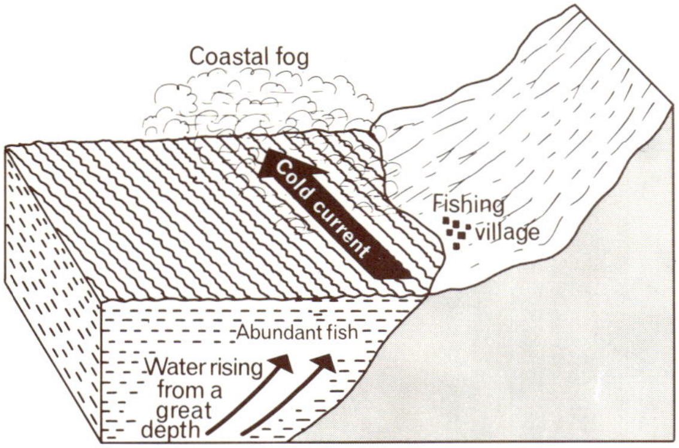

Fig. 35 Cold current and upwelling water

damp air rises and cools rapidly at the coastal mountains where considerable precipitation takes place. When this happens disaster hits the coastal peoples. Torrential rain causes severe soil erosion and flooding on the dry dusty land. Off shore, fish die by the thousand.

Another result of the pattern of warm and cold currents in the tropics is seen in the distribution of coral. It requires a sea temperature of at least 20°C (68°F) and is much more abundant on the east side of continents than on the west (e.g. the Great Barrier Reef of eastern Australia). In certain areas of powerful warm currents, coral reefs are found well beyond the tropics. A good example of this is Bermuda where, because of the Gulf Stream, there are coral reefs at 32°N.

The influence of ocean currents outwith the tropics is well seen in the North Atlantic. There the Gulf Stream has a speed of up to 4 knots and may be 10°C warmer than the surrounding ocean. The marked temperature change in its vicinity has a considerable influence on the weather, and it is a noted area for storms off the coast of America. Further north, where it swings away from the Canadian coast near the Newfoundland banks, it comes in contact with the cold Labrador current producing an area notorious for its sea fogs (fig. 36). In the same region it is responsible for the rapid melting of icebergs brought from Greenland by the cold currents.

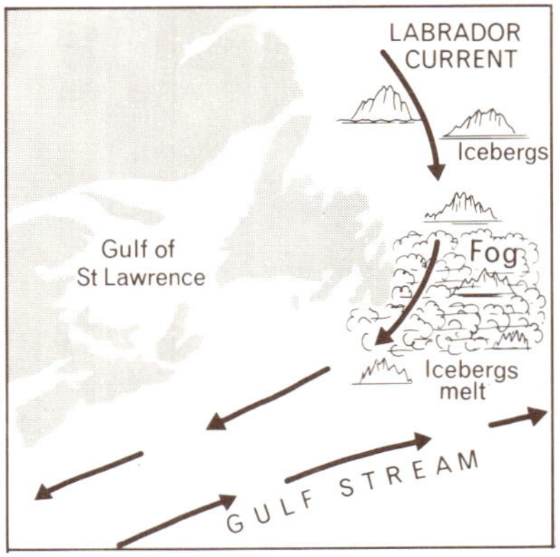

Fig. 36 Cold and warm current meeting

The most outstanding effect of currents on temperatures of the North Atlantic coastlands is seen when conditions on the eastern and western seaboards are compared as in the following table and in fig. 37.

Town	*Average Temperatures* Jan.	July	
Brest (48°N) (France)	6·7°C	17·2°C	Brest is 11·7°C warmer than St. John's in January and 2·2°C warmer in July.
St. John's (47°N) (Newfoundland)	−5·0°C	15·0°C	
Stavanger (59°N) (Norway)	1·1°C	15·0°C	Stavanger is 22·1°C warmer than Hebron in January and 6·7°C warmer in July.
Hebron (58°N) (Canada)	−21°C	8·3°C	

These differences are largely brought about by the influence of the Labrador current on the Canadian coast, and the North Atlantic drift on the European coast. Although they are most marked in winter, they are still noticeable in summer. This unusually

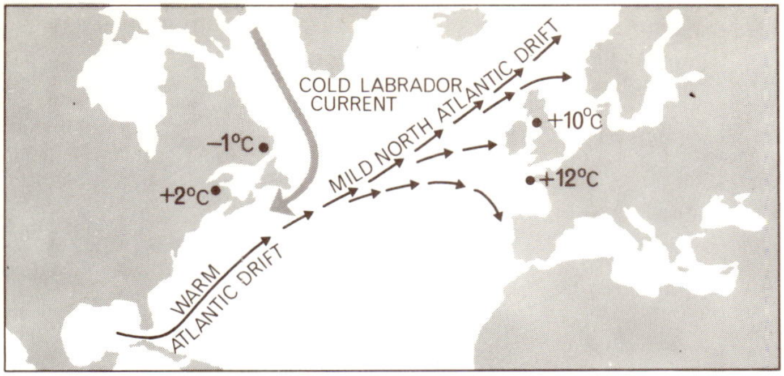

Fig. 37 Differences in temperature between east and west coasts of the Atlantic on 26th November 1966

high temperature for the latitude which occurs in north-west Europe is known as a **positive temperature anomaly**, and it has important effects on the development of the coastal regions. Icebergs and the closure of parts by frozen sea are unknown in Norway, where small fishing and farming communities are found beyond 70°N. In contrast, coastal Labrador, 1600 km (1000 miles) to the south, is almost uninhabited and farming is virtually impossible.

STUDY SECTION ON CHAPTER 3

1 What is the difference between an *ocean current* and an *ocean drift*?

2 Why is the St. Lawrence river closed to navigation for part of the year while the Clyde is never closed?

3 In what direction do cold currents flow?

4 Identify the currents shown on the map and state which are warm and which are cold.

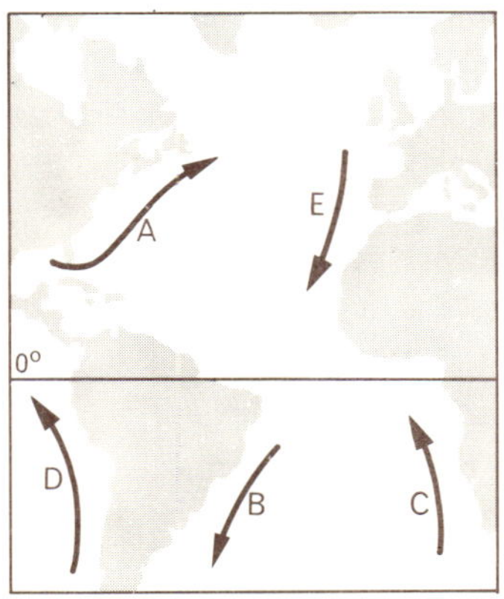

5 Why do the main ocean currents have a clockwise movement in the northern hemisphere and an anticlockwise movement in the southern hemisphere?

Water Vapour

4

In the introduction it was stated that one of the gases in the atmosphere is called *water vapour* and that its proportion is always varying as some of it may be in the form of a liquid (water) or a solid (ice). These changes in state are mainly caused by changes in temperature, and this can be illustrated by warming a block of ice in a pan as shown in fig. 38. It should be

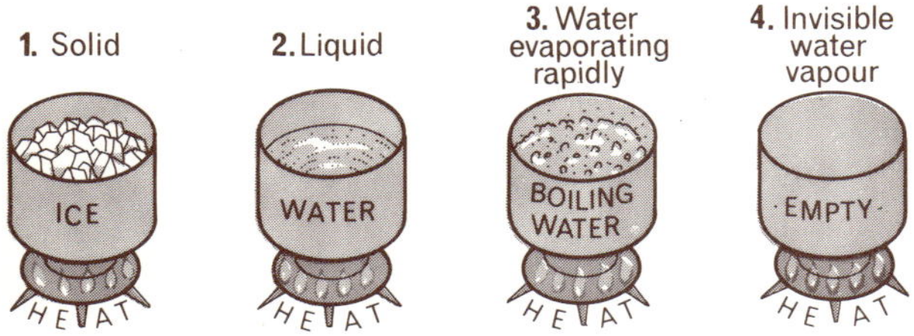

Fig. 38 Heat changes solid water (ice) first into a liquid and then into a gas

noted that (*a*) liquid water can change into water vapour without being boiled, but boiling speeds up the process; and (*b*) water vapour is invisible. The steam we see coming from a pan of boiling water is made up of tiny drops of liquid and is not water vapour.

Humidity

The amount of water vapour present in a given volume of air is called its **absolute humidity**. When this air moves over oceans, lakes, and rivers, it receives more water vapour by evaporation. There is a limit to the amount of water vapour it can hold and, when this limit is reached, the air is said to be **saturated** or to have a **relative humidity** of 100%.

The capacity of air to hold water vapour is increased if the temperature rises. Thus if saturated air is heated without the addition of any more water vapour (i.e. the absolute humidity remains the same) it ceases to be saturated and its relative humidity drops below 100%.

If, on the other hand, saturated air is cooled, its capacity to hold water vapour is decreased, and the surplus vapour **condenses** as small drops of water. In this case the relative humidity remains at 100% but the absolute humidity decreases. The temperature at

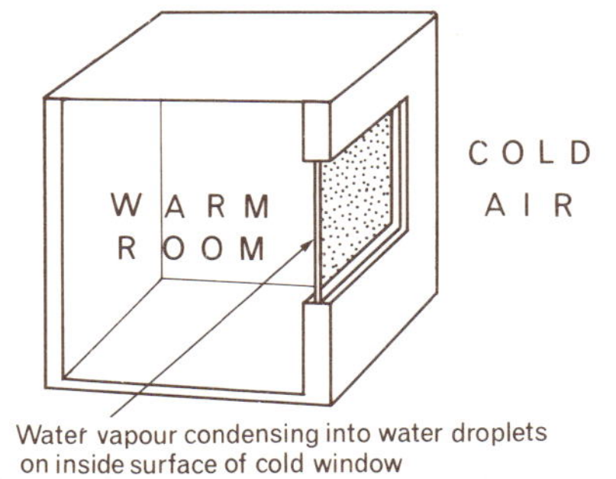

Fig. 39 Water vapour condenses when air temperature is lowered

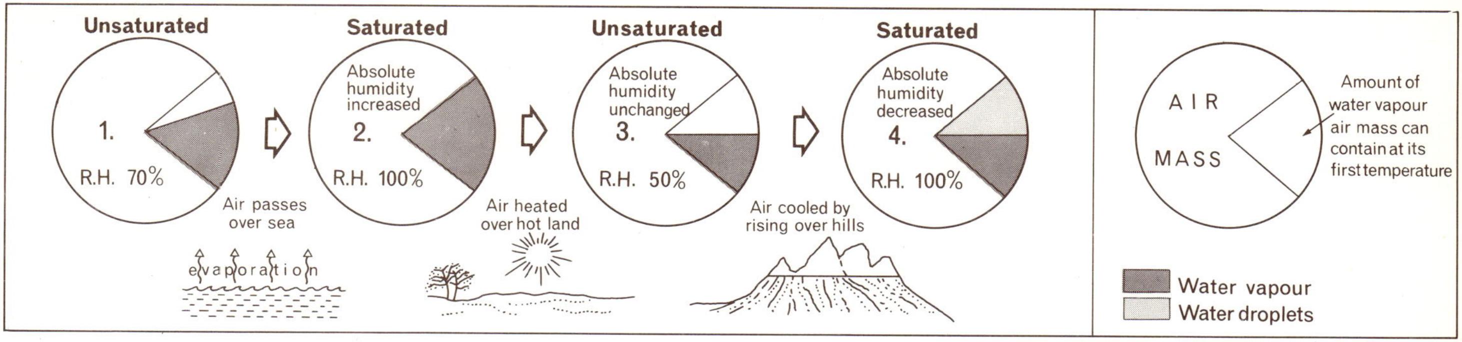

Fig. 40 Changes in the humidity of a small air mass as it moves over the earth

which condensation starts is called the **dew point**. These changes are illustrated in figs. 39 and 40.

Latent Heat

When water reaches boiling-point, further heating will not raise its temperature any higher. Instead the heat will be used to cause evaporation of the water. Heat used in this way to change liquid into gas is called **latent heat**. It is used whenever water is evaporating, whether it is boiling or not. This process also works in reverse: when water vapour is changing into water droplets, and condensation is taking place, latent heat is released into the atmosphere.

The effect of latent heat on the human body is felt by a swimmer. He feels colder when he comes out of the water than he did before he went in, although the air temperature is the same. This difference is caused by latent heat evaporating water from his skin. It is only when this evaporation is complete that the heat from the sun starts to warm him up (fig. 41).

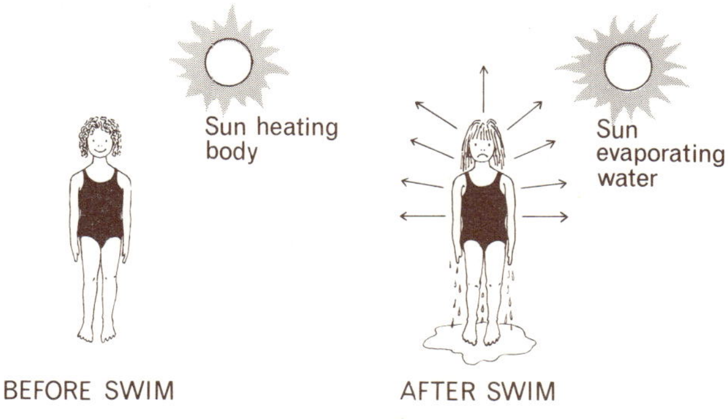

Fig. 41 A damp body is cooler than a dry one when the sun's heat is being used to evaporate water

Measurement of Relative Humidity

The difference in temperature between a wet and a dry body is used to measure the relative humidity of

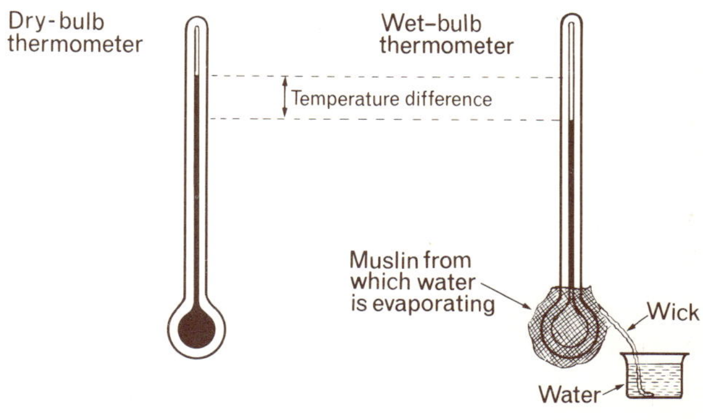

Fig. 42 Wet-and-dry-bulb thermometers

the atmosphere. Two identical thermometers are housed beside each other in a Stevenson screen (see page 38), and one of them has a piece of damp muslin tied round the bulb (fig. 42). Evaporation from this muslin keeps the temperature of the **wet-bulb** thermometer below that of the **dry-bulb** thermometer. When the air is dry (relative humidity well below 100%) evaporation is rapid and there is a large difference in temperature between the two thermometers. When the air is damp (relative humidity near 100%) evaporation is slight and the temperature difference is much less. A set of tables is used to convert the temperature difference into a measurement of relative humidity as a percentage. Clothes on a washing line dry much more quickly on a day when the relative humidity is low, than on a day when it is high.

Various Forms of Condensation

It has already been stated that all air contains some water vapour. The amount varies greatly from place to place: air which has blown over an ocean is much damper (higher absolute humidity) than air which has blown over a large continent. Before any water vapour will condense into water droplets, the temperature must drop sufficiently for the dew point to be reached. This point will be arrived at more readily in damp air than in dry air. This is the reason why rainfall is much more common in maritime regions than in the interior of continents.

Condensation takes a number of forms and the main ones are cloud, rain, snow, fog, dew and frost.

Clouds

Dry air rising in the troposphere expands and cools by a process called **adiabatic cooling**. The presence of water vapour complicates this and the average rate at which the temperature drops is 0·6°C for every 100 metres (1°F per 300 feet). This drop in temperature is known as the **lapse rate**. When the temperature of ascending air reaches its dew point condensation takes place and the water droplets form clouds some distance above the earth's surface. The height at which this takes place is called the **condensation level** (fig. 43). Although the tops of a group of clouds

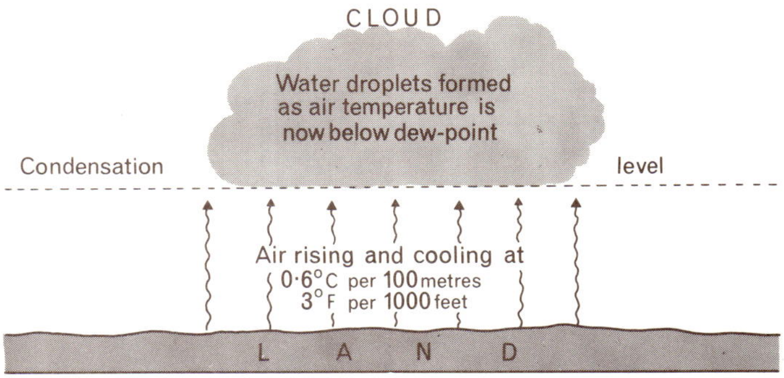

Fig. 43 Clouds forming as a result of air rising

may vary greatly in height, they normally all have their bases at the same level. This is sometimes called the **cloud ceiling** or **cloud base**.

Rising currents of air result from three basic causes, each of which tends to produce a different type of cloud.

1 OROGRAPHIC OR RELIEF CLOUD

Orographic or relief cloud results from air rising over a hill or mountain. It produces a cloud cap which tends to be restricted to the immediate neighbourhood of the upland. On the leeward side of the upland, the condensation level and cloud base is normally higher than on the windward side, as the humidity of the air has been reduced by condensation in the rising air. This higher cloud base, drier air, and descending air result in the leeward side of uplands being much drier and giving a **rain shadow** area (fig. 44).

Fig. 44 Formation of orographic cloud

2 CONVECTION CLOUDS

Convection clouds are formed when convection currents of air rise from a hot land surface. These clouds extend higher than orographic clouds. They are most common in the afternoon when the land is at its warmest. They are a characteristic feature of tropical climates where they are frequently associated with the development of thunderstorms (fig. 45).

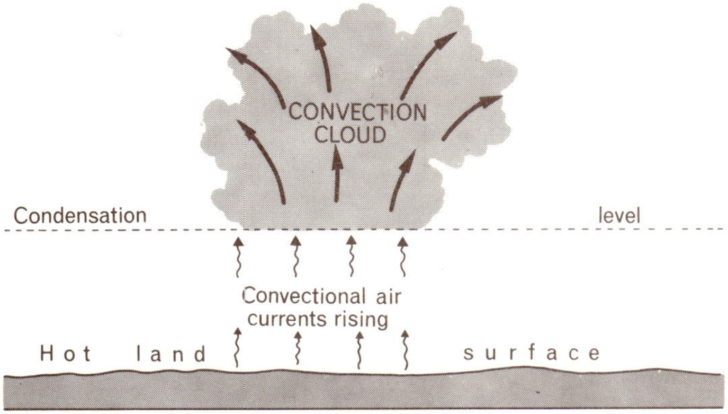

Fig. 45 Formation of convection cloud

3 FRONTAL CLOUDS

In fig. 26 it was shown that when two contrasting air masses meet, air currents rise along a boundary called a *front*. Clouds form along these fronts because the air cools as it rises, and also because the two bodies of air are at different temperatures and one is cooled by the other (fig. 46). Clouds of this type are often much more extensive than those formed by uplands or by convection currents and are extremely common in the British Isles.

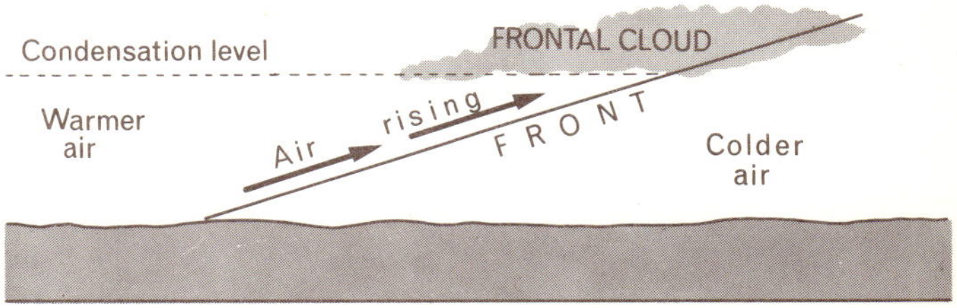

Fig. 46 Formation of frontal cloud

Stable and Unstable Air

The upward movement of air which leads to the formation of clouds is started by one or more of the basic factors mentioned in the previous paragraphs, but the height to which these currents rise is largely decided by whether the air is stable or not. In Chapter 5 it will be seen that the temperature of the air in the troposphere decreases with altitude, therefore the rising currents will be ascending through air which is becoming steadily colder. If the drop in temperature of the surrounding air is more than 3°C per 300 metres—the *lapse rate* of the dry ascending air—the currents will continue to rise after the factor which started them off has ceased to have any effect (because the rising air will always be warmer than the surrounding air). When this happens conditions are **unstable**. On the other hand, if the temperature drop in the surrounding air is less than 3°C per 300 metres, the upcurrents will be cooler than the surrounding air and will stop once they cease to be affected by their starting factors. Such conditions are **stable**. During unstable conditions clouds are much higher and more widespread than when the atmosphere is stable. The differences are shown in figs. 47 and 48.

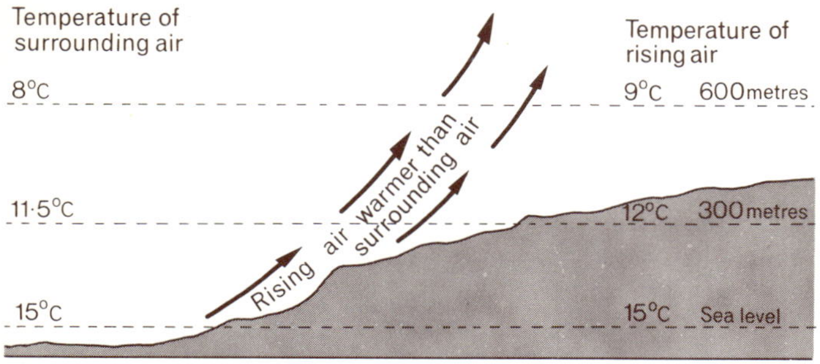

Fig. 47 Unstable air
Air is forced to rise by the hill, and it continues to rise above the hill as it is always warmer than the surrounding air. These conditions often lead to the formation of large clouds, especially if the relative humidity is high.

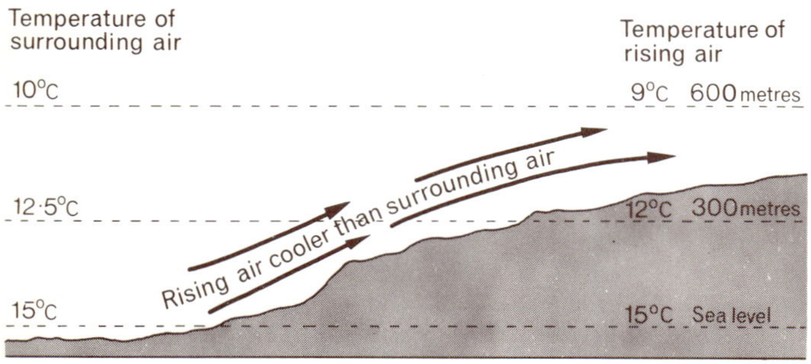

Fig. 48 Stable air
Air is forced to rise by the hill, but the upward movement ceases when it reaches the summit as it is cooler than the surrounding air. A low blanket of cloud will form on the hill if condensation level is below the top of the hill.

PRECIPITATION

Precipitation is the name given to the rain, hail or snow which sometimes falls from clouds. As soon as water droplets are formed in a cloud they start to fall (fig. 49). Their speed depends on the size of the droplets, with the larger ones falling more rapidly than the smaller ones. This is why the small droplets in a drizzle land gently on the ground, while the large droplets from a thunder cloud bounce off the roads. When droplets (or **rain**) start falling from a cloud, they are falling into warmer air, and many of them evaporate before they reach the ground. There may also be strong upcurrents of air which keep the droplets aloft. These two factors, combined with the size

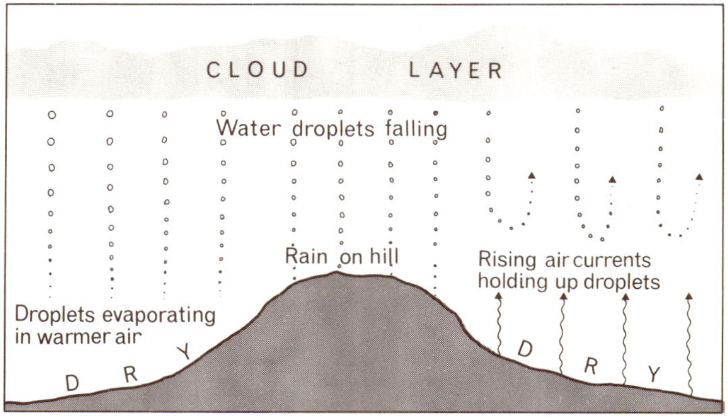

Fig. 49 Hills may be wet, while bordering lowlands are dry

of the droplets and the altitude of the cloud, decide whether or not rain will reach the ground. In high clouds, formed by strong upcurrents, some of the droplets may be carried so high up that they freeze and change into small balls of ice which fall to the ground as **hail** (fig. 50). These conditions often occur during hot weather (causing strong convection currents) when the air is unstable. **Thunder** and **lightning** may accompany the hail storm.

In winter when the temperature of the air is below freezing-point (0°C) almost to ground level, the water vapour condenses as small crystals of ice called **snow** (fig. 51). (This should not be confused with hail which is frozen rain.) If the temperature at ground level is just above freezing, the snow may partly melt before it lands; in this state it is known as **sleet**.

The great variety of factors, such as stability of the air, height of condensation and freezing levels, and strength of upcurrents, affecting the formation of clouds, result in many different types being formed; some of them are illustrated in fig. 52.

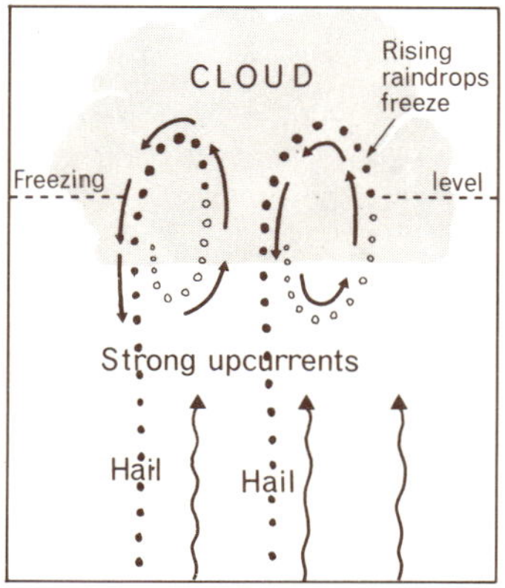

Fig. 50 Formation of hail

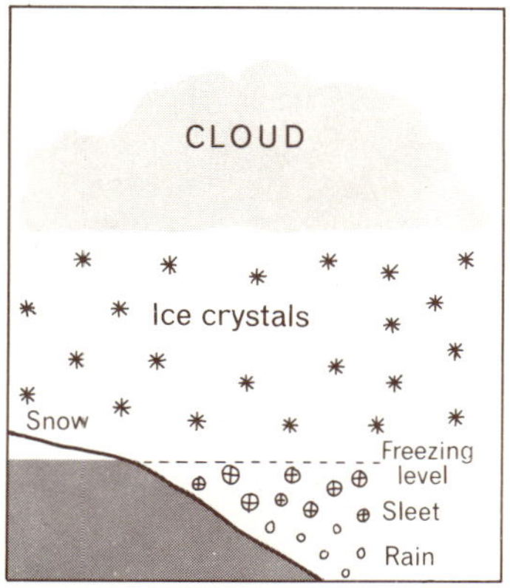

Fig. 51 Formation of snow

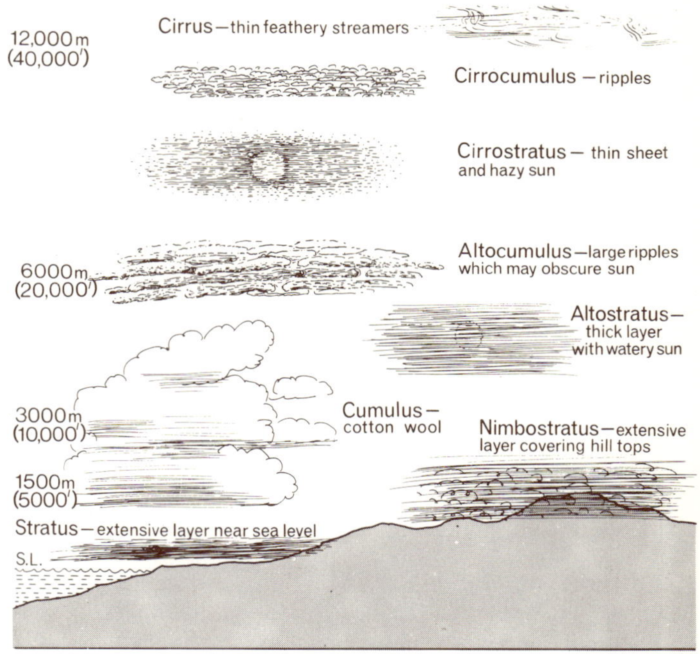

Fig. 52 Some common cloud types

FOG

Fog is another form of condensation, but its formation differs from that of clouds:

1. No upcurrents are involved, and in some cases the air may be actually descending.
2. The cooling of the air to its dew point is caused by the air coming in contact with cold land or water.
3. The water droplets are formed in the layer of air in immediate contact with the surface of the earth.

The water droplets which condense from the air to form fog greatly restrict visibility. When there is a large amount of soot and dirt in the atmosphere, it sticks to the water, giving a very dense unpleasant form of fog, sometimes called **smog**.

True fog should not be confused with **hill fog**

which is simply very low cloud covering the summits.

Two main types of fog can be distinguished by the way in which the process of cooling the air takes place.

RADIATION FOG

Radiation fog occurs during calm cloudless nights when the surface of the ground cools rapidly by radiation of heat into the atmosphere (fig. 53). This

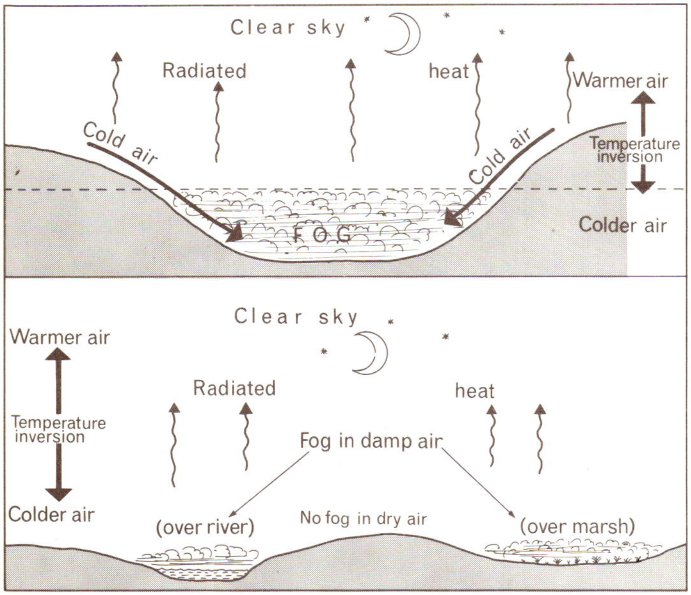

Fig. 53 Conditions for the formation of radiation fog
Fog disperses during the day if the sun raises the temperature high enough for the water droplets to evaporate, or if the wind breaks up the temperature inversion.

causes a **temperature inversion** in the lowest layers of the atmosphere, with the air nearest to the ground colder than the air immediately above it. If the cold air has a fairly high relative humidity, it will soon reach its dew point; water droplets will condense to form fog.

Fog is unlikely to occur if:

(*a*) there is cloud cover which would stop the land losing its heat by radiation.

(*b*) the relative humidity is low.

(*c*) there is a wind stronger than about 16 k.p.h. (10 m.p.h.) which would cause movement of the air and destroy the cold layer.

When conditions are suitable for the formation of fog, it is most common in valleys and hollows. There the frequent occurrence of rivers, lakes and marshes locally raises the relative humidity, and also cold air tends to drain downhill.

ADVECTION FOG

Advection fog occurs in moving air when a mass of mild damp air blows across cold sea or land. The air in contact with the cold surface is cooled to its dew point and fog forms. Two situations in which this can occur are illustrated in fig. 54.

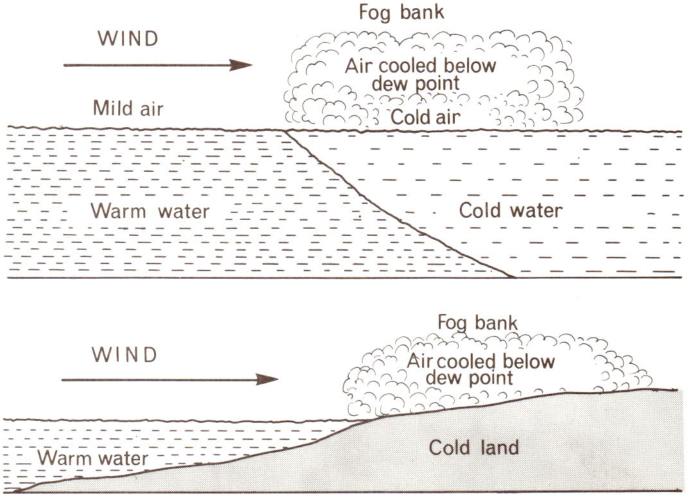

Fig. 54 Formation of sea and coastal (advection) fog

DEW

Dew is also formed during calm clear nights when the ground loses its heat by radiation (fig. 55). Now it is only the air in actual contact with the ground

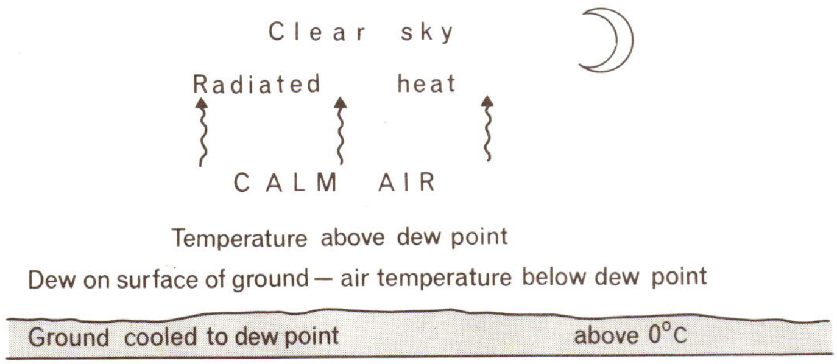

Fig. 55 Formation of dew

which is cooled to its dew point, and the water droplets form on the earth's surface. Dew is most common on grass and other plants, as their moisture locally increases the humidity of the air.

FROST

Frost forms in the same conditions as dew (fig. 56). This time the dew point is below freezing-point and the water is precipitated as needle-like ice crystals or **hoar frost**. As with dew it is commonly seen on plants.

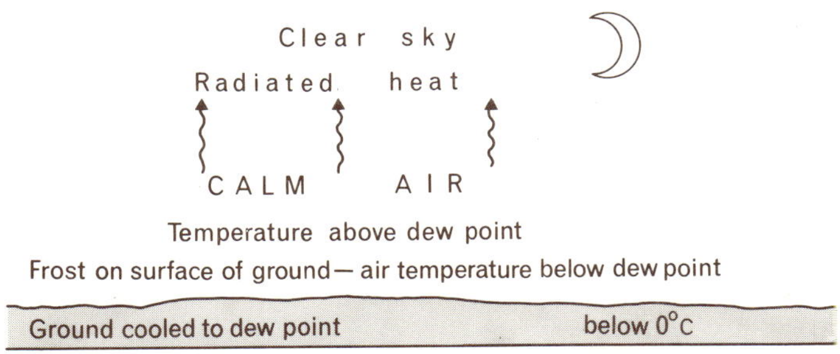

Fig. 56 Formation of hoar frost

Another form of frost called **glazed frost** is formed when rain falls on a ground surface which is below freezing-point, giving a layer of ice (unlike the white crystals of hoar frost). This glazed frost is extremely dangerous when it forms on road surfaces or when sea-spray freezes on the superstructures of trawlers in the Arctic, making them unstable and liable to capsize.

Measurement of Rainfall

Rainfall is measured simply by collecting rain in a rain gauge (fig. 57). The container is emptied into a graduated measuring cylinder at fixed times, and the depth of water is measured in millimetres or inches.

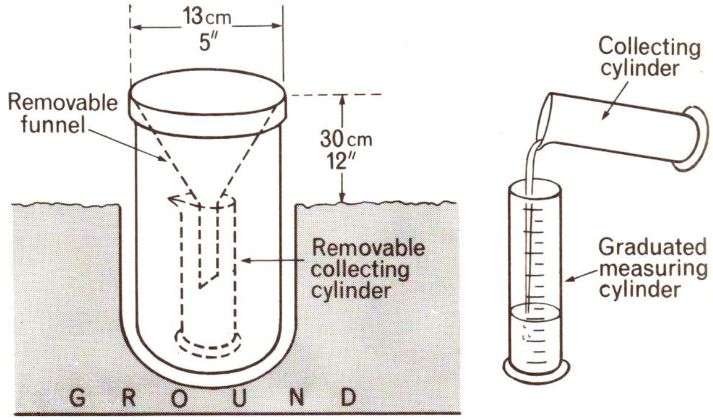

Fig. 57 Rain gauge

Hail, sleet, and snow have to be melted before being measured, but a rough calculation can be made by taking 10 mm snow equal to 1 mm rain (1 foot of snow as equal to 1 inch of rain, as it is a very rough rule).

RECORDING OF RAINFALL

Once regular daily rainfall measurements have been made over a number of years, averages for each month and the average annual rainfall can be calculated. These averages can then be shown on maps and graphs.

To make a rainfall map the averages for each station are plotted. By interpolation, lines of equal rainfall, called **isohyets**, can be drawn in the same way as isobars are drawn on a pressure map (see fig. 7). Once the isohyets have been drawn, the spaces between the lines are shaded (or coloured), darker for the wet areas and lighter for the dry areas as shown in fig. 58.

A map is ideal for showing variations in rainfall from place to place; but the variations at any one place from month to month—the **seasonal distribution** or **rainfall regime**—are best shown on a bar graph as illustrated in fig. 59. The Winnipeg graph shows that, while every month has some rain, there is a well-marked wet season from May to September. In the Halifax bar graph, there is a considerable variation in rainfall from month to month, but there is no well-defined wet or dry season. Halifax could be described as a place with rain at all seasons, while Winnipeg has rain mainly in summer. Using this type of information for a large variety of places, we find that, as well as being divided into regions of heavy, moderate, and light rainfall (fig. 58), the world has a pattern of seasonal rainfall which is illustrated in fig. 60.

Both the amount and the seasonal distribution of

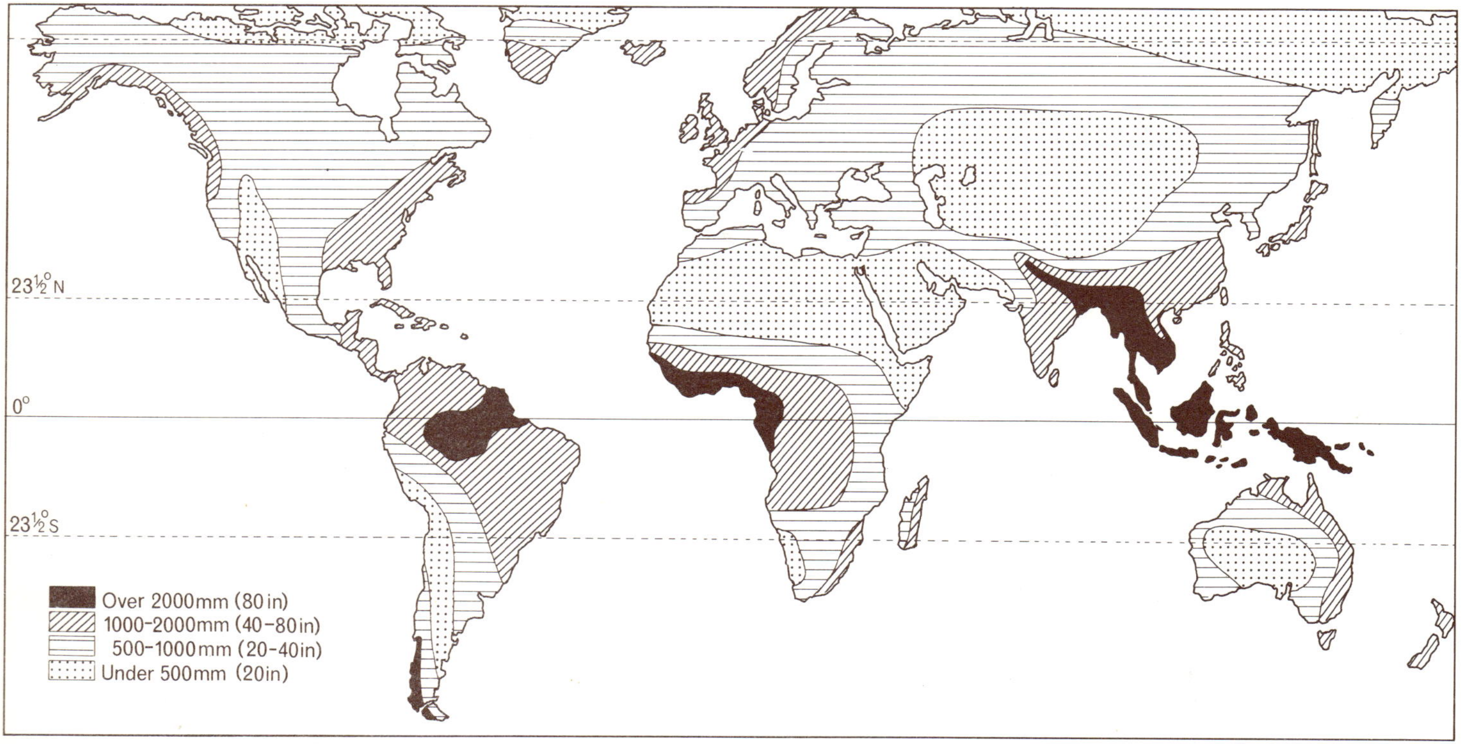

Fig. 58 Simplified annual rainfall map of the world

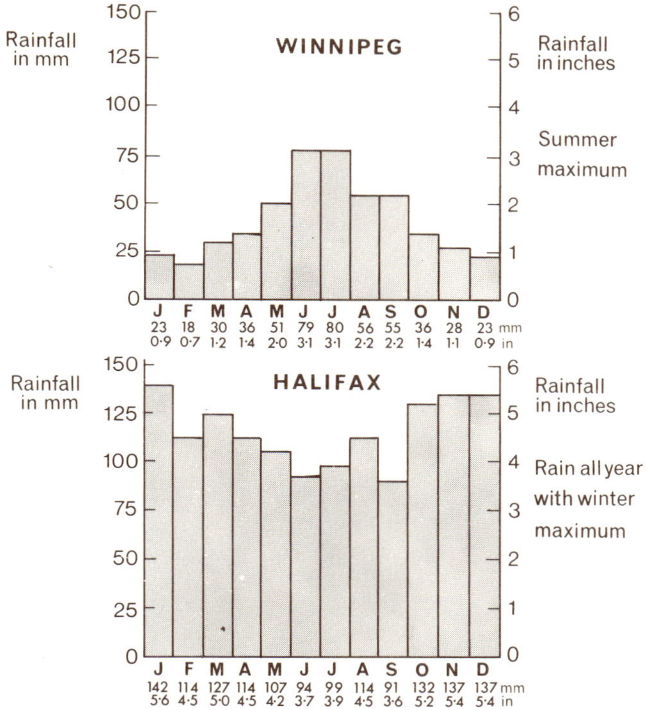

Fig. 59 Bar graphs illustrating the seasonal distribution of rainfall

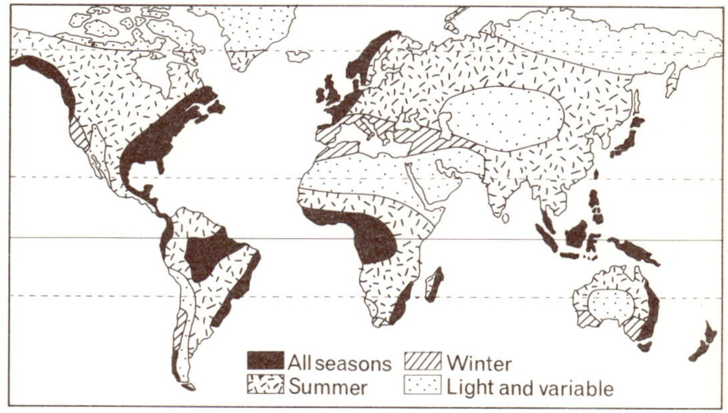

Fig. 60 Seasonal distribution of rain

rainfall are very important to man, as water is essential for his existence. In areas of very light rainfall, such as Egypt, cultivation is impossible without irrigation. Britain normally has ample rain throughout the year to meet all farming requirements. Temperatures can have a considerable influence on the effectiveness of the rainfall. In the tropics 500 mm (20 inches) may be insufficient for normal cultivation due to the high evaporation rate, whereas on the Canadian prairies it would be adequate—partly because of the lower evaporation rate, and partly because the winter precipitation comes in the form of snow which is available for crops when it thaws in the spring. Monsoon countries like India illustrate clearly the importance of seasonal distribution: although temperatures are high enough for cultivation throughout the year, crops can be grown in winter (the dry season) only if some of the summer (wet season) water has been stored for irrigation.

It is not only for agriculture that the rainfall pattern is important. In a highly urbanized industrial country like Britain, water is required for domestic water supply, industry, and hydro-electric power stations. Areas of heavy rainfall, such as the Welsh mountains, the Lake District, and the Scottish Highlands, supply much of the water used in the industrial towns of the adjacent lowlands, where the rainfall is insufficient to meet all the requirements.

STUDY SECTION ON CHAPTER 4

1 What is the difference between *absolute humidity* and *relative humidity*?

2 What is the relative humidity of saturated air?

3 Why do water droplets form when saturated air is cooled?

4 What name is given to the temperature at which condensation starts?

5 Copy the diagram of wet- and dry-bulb thermometers in fig. 42, and explain how these thermometers measure relative humidity.

6 Why are coastal areas often wetter than the interiors of continents?

7 What fixes the height at which cloud base occurs?

8 Explain why the leeward side of a hill often has a higher cloud base and less rainfall than the windward side.

9 What do the terms (*a*) unstable air and (*b*) stable air mean?

10 How does snow differ from hail in its method of formation?

11 Describe with the aid of a diagram how radiation fog is formed.

12 What is a temperature inversion?

13 Describe the conditions under which frost may occur.

14 How is mean annual rainfall measured and calculated?

15 Use climate statistics or climate maps to find (*a*) mean annual rainfall, (*b*) wet season (if any) for each of the following places: Bergen, Moscow, Rome, Santiago, Fremantle, and Calcutta.

Temperature

5

The sun is the source of all the heat in the atmosphere. The heat travels through space to the earth in the form of short waves called **solar radiation** or **insolation**. These short waves pass through the atmosphere to the earth's surface, and in doing so they lose a certain amount of heat from absorption by water droplets and dust particles. The angle at which they strike the earth's surface is very important in deciding how much heat the earth receives, and this is illustrated in figs. 61–63.

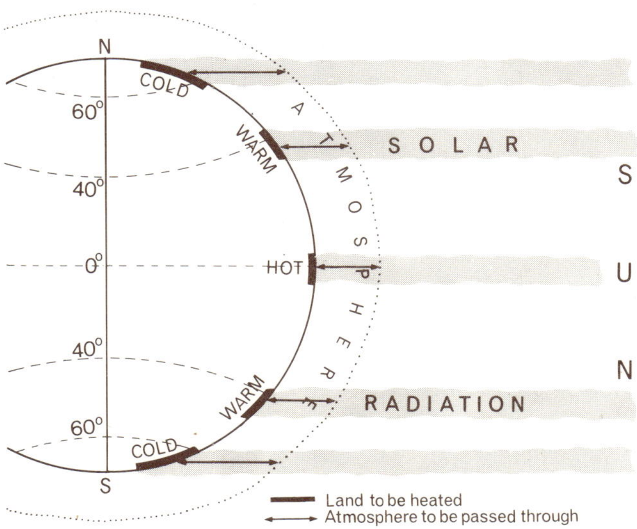

Fig. 61 Temperature decreases from the equator to the poles
Equal rays of solar radiation (heat) have to pass through more atmosphere and heat more land at the poles than at the equator

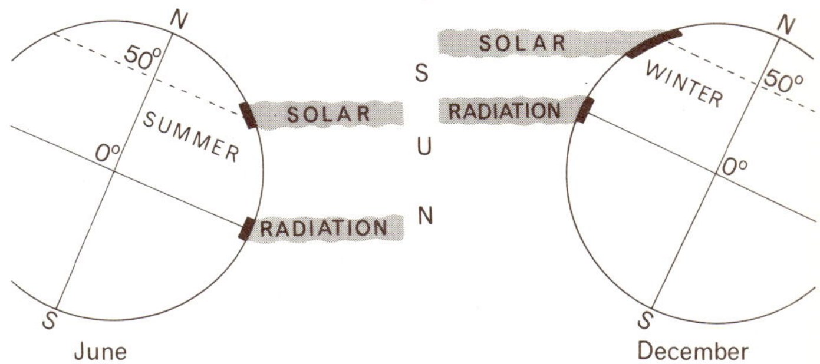

Fig. 62 Why summer is warmer than winter
At 50° N equal rays of solar radiation (heat) have to heat a larger part of the earth's surface in winter than in summer. At the equator the area to be heated is similar all the year. As a result, in regions outside the tropics it is much colder in winter than in summer; within the tropics, temperatures vary only slightly during the year.

Although the atmosphere receives a little heat from the short waves of solar radiation, most of the heating is done by long waves which come from the earth (fig. 64). These waves heat the air by **radiation** (as from an electric radiator), and **convection** (as from an electric convector heater), and **conduction** (as from the hot plate of an electric cooker). The tem-

Fig. 63 Why one side of a valley is warmer than the other
The side of a valley facing the sun is receiving more concentrated solar radiation (heat) than the opposite side, although the sun is shining on both sides.

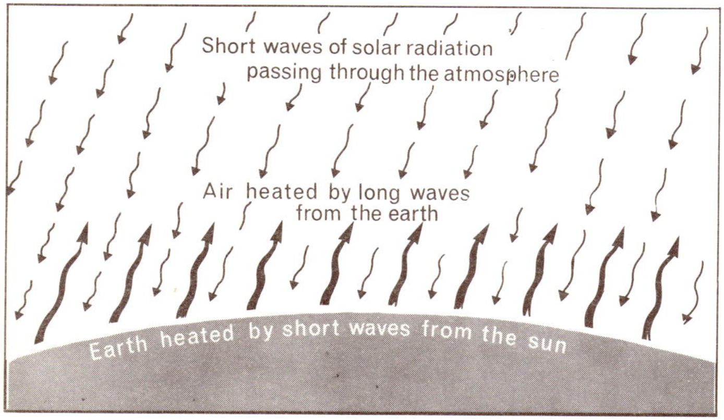

Fig. 64 The atmosphere is heated by long waves from the earth

perature of the air depends on how much the earth below it has been heated by the sun. Clearly the air at the equator will be warmer than the air at the poles (fig. 61), and in temperate latitudes (e.g. 50°N) the air will be warmer in summer than in winter (fig. 62).

The distribution of **land** and **sea** also has an effect on air temperature, as the sea heats up more slowly than the land in summer. It also cools down more slowly in the winter, so that sea or **maritime air** tends to be mild all year, while **continental air** is hot in summer and cold in winter (fig. 65).

The temperature of the troposphere also decreases with height or **altitude** at a rate of about 0·6°C per 100 metres. This is because there is less land to heat the air at high altitudes (fig. 66) and most of the heat from the earth is used to warm the air immediately above it.

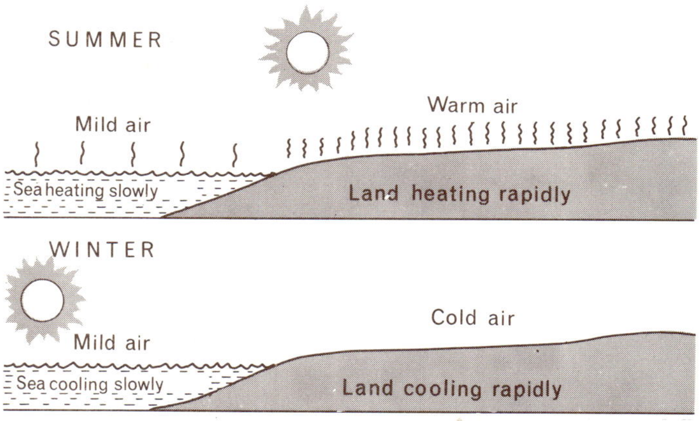

Fig. 65 Land and sea affect air temperature

Measurement of Temperature

The temperature of the air is measured by a **maximum-and-minimum thermometer** (fig. 67) which gives the highest and lowest temperatures reached during the day. This instrument is housed in a wooden box called a **Stevenson screen** (fig. 68) which also contains the wet-and-dry-bulb thermometer. The slots in the screen allow free passage of air around the thermometers. They keep out direct sunlight, and these **shade temperatures** may be considerably lower than those measured at places where the sun is shining on a thermometer. An additional minimum thermometer is placed about 25 mm (1 in) above the ground over grass, and is used to measure the minimum night temperature.

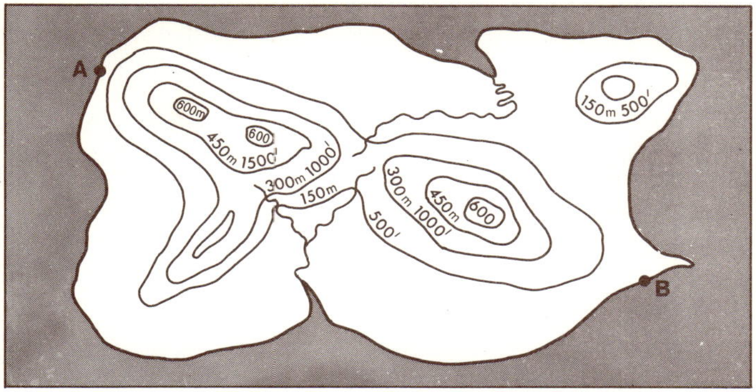

(A) Map of an island (vertical interval 150 metres, 500 feet)

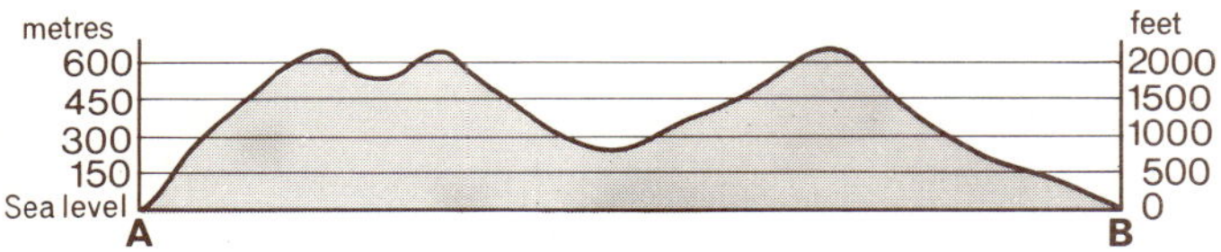

(B) Cross section of an island from **A** to **B**

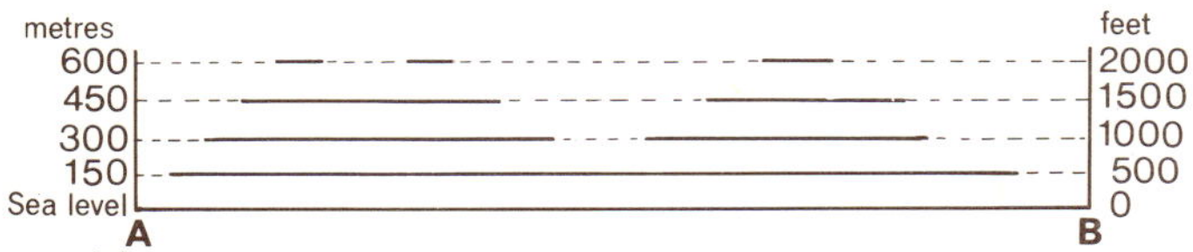

(C) Amount of land at each contour along **AB**

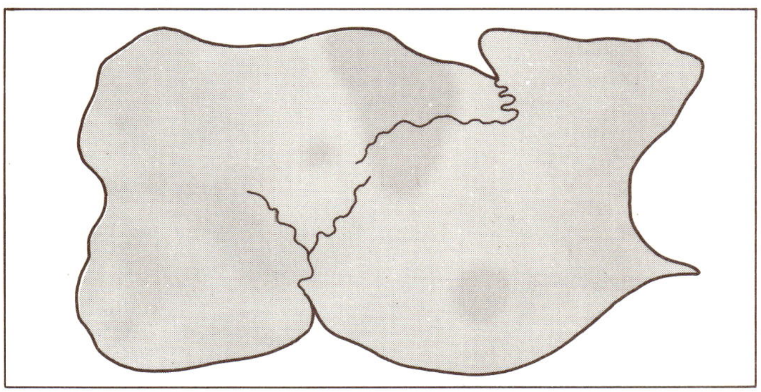

(D) Area of land at sea level

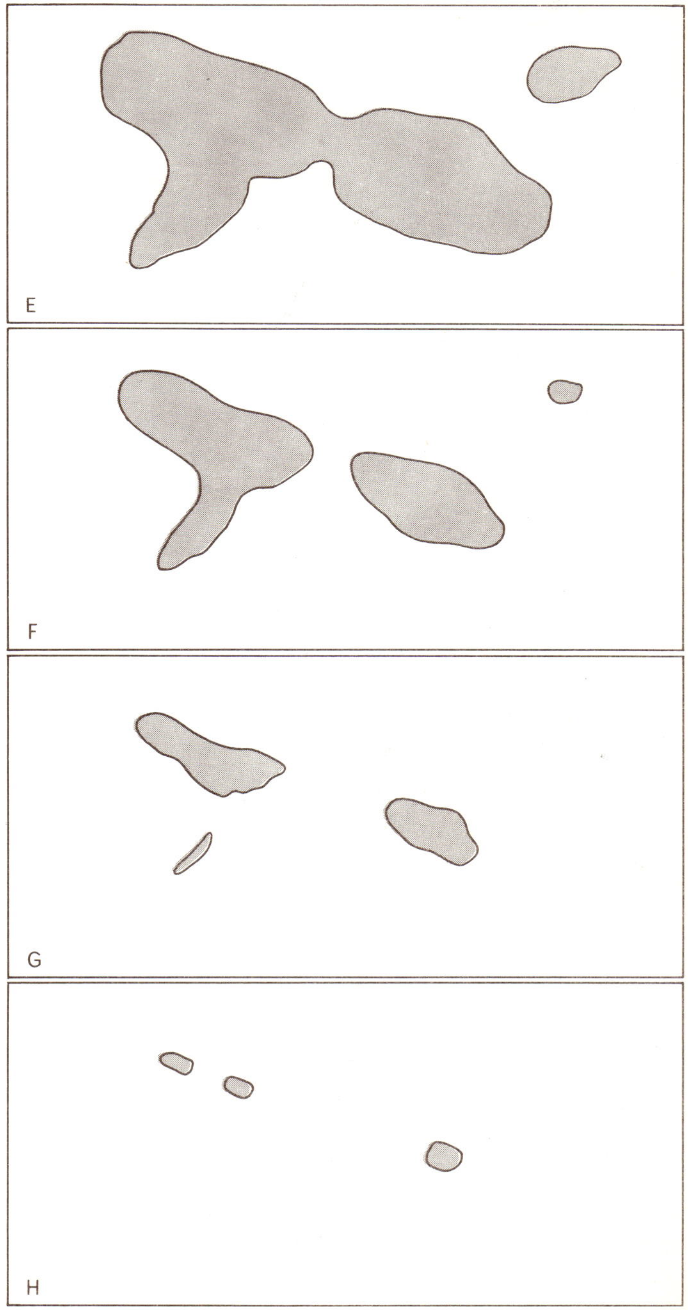

(E) Area of land at 150 metres, 500 feet
(F) Area of land at 300 metres, 1000 feet
(G) Area of land at 450 metres, 1500 feet
(H) Area of land at 600 metres, 2000 feet

Fig. 66 Land area decreases with altitude

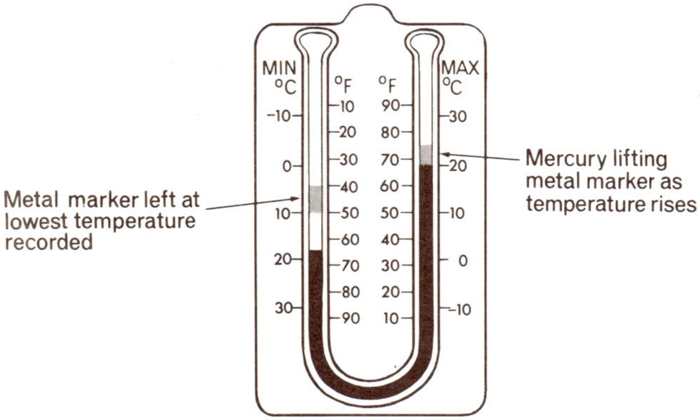

Fig. 67 Maximum-and-minimum thermometer

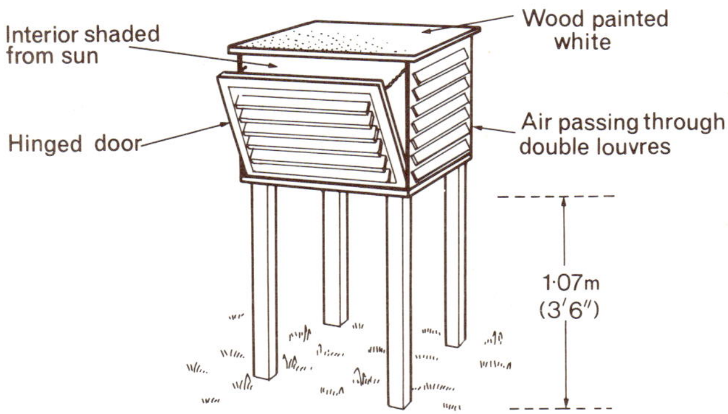

Fig. 68 Stevenson screen

TEMPERATURE MAPS AND GRAPHS

Temperature maps and graphs are drawn from average (or mean) temperatures calculated from the daily measurements. As with rainfall, a map is the best means of showing variations in temperature from place to place, while line graphs show the seasonal variations at one particular place.

Maps are drawn by plotting lines of equal temperature (called **isotherms**) in the same way as shown for isobars on fig. 7. The spaces between the lines are then shaded or coloured according to a suitable scale. On many temperature maps a second set of isotherms is drawn using temperatures reduced to sea-level (fig. 69). This is done by adding 0·6°C for every 100 metres that the measuring station is above sea-level. These sea-level isotherms are very

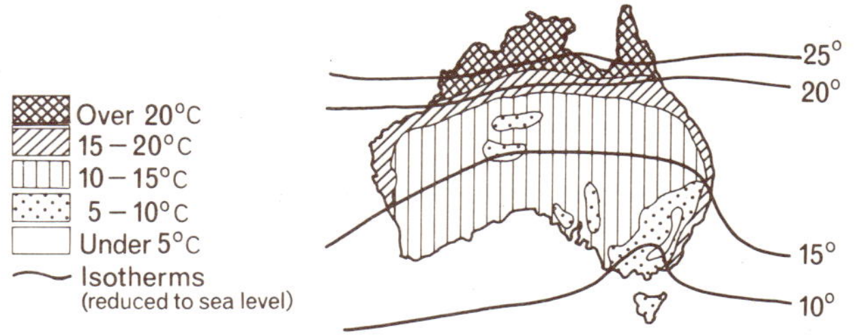

Fig. 69 Map showing July temperatures in Australia

useful for showing the way in which the temperature pattern is influenced by latitude and the land-sea distributions. Temperature cannot be collected like rainfall, so it is normally shown on curved-line graphs (fig. 70), whereas monthly accumulations of rain are best shown on bar graphs (fig. 59).

The difference in temperature between the warmest and coldest months is called the **annual temperature**

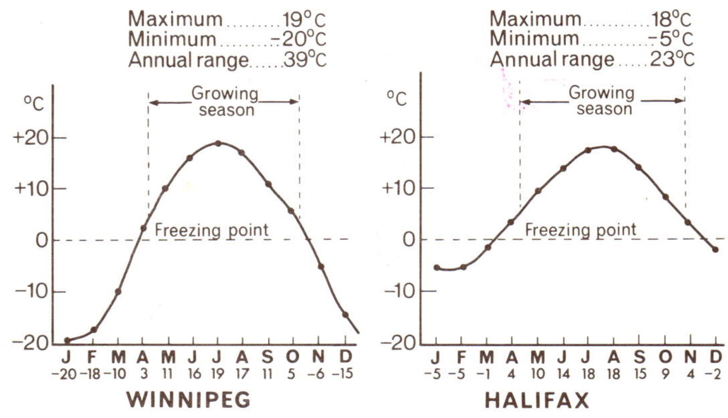

Fig. 70 Temperature graphs for Winnipeg and Halifax

range. It can be obtained easily from the mean monthly figures. The mean monthly temperatures which are normally given in atlases and books do not show the temperature difference between day and night (the **diurnal range**) however. This information is of great importance to farmers, particularly in the spring, when night frost can cause serious damage to certain crops. Desert areas with their clear skies have a large diurnal range, and night frost is quite common during winter in the tropical deserts.

Another very important temperature is 5·5°C as it marks the limits of the **growing season**. Although many plants can survive temperatures below freezing, it is only when the monthly average is above 5·5°C that significant growth takes place. It will be seen in fig. 70 that, although Winnipeg has a much colder winter than Halifax, there is very little difference in the lengths of the growing seasons.

THE WORLD TEMPERATURE PATTERN

The world temperature pattern is shown in fig. 71. The following features should be particularly noted:

1. The isotherms have an east–west trend emphasizing the effect distance from the equator has on world temperatures.
2. The temperature belts migrate north and south with the overhead sun.
3. The area between the tropics seldom has temperatures below 20°C (except on mountains).
4. Outside the tropics there are well-marked hot (summer) and cold (winter) seasons. This does not show up very well in the southern hemisphere because of the limited amount of land beyond 30°S.

On the basis of these maps it is convenient to divide the world into **temperature regions** or **zones** as shown on fig. 72.

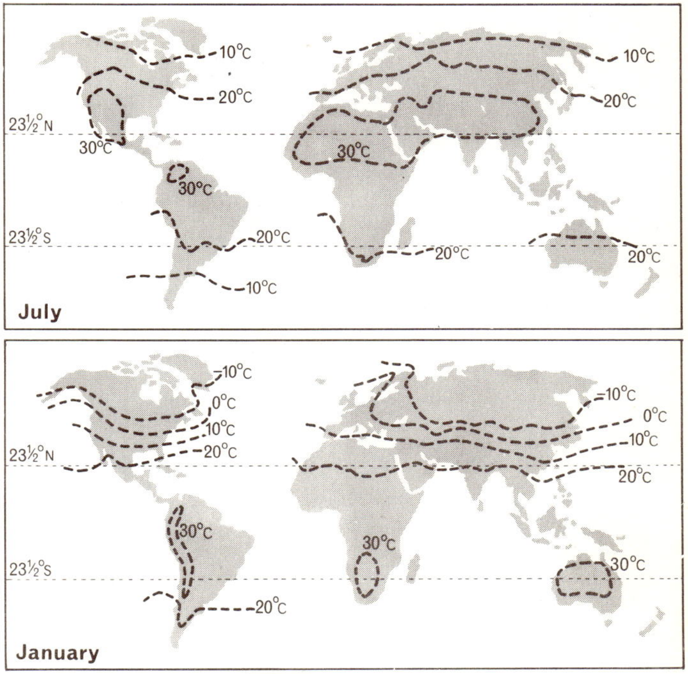

Fig. 71 World isotherms (simplified)
These are sea-level isotherms, so mountains do not stand out as cold areas.

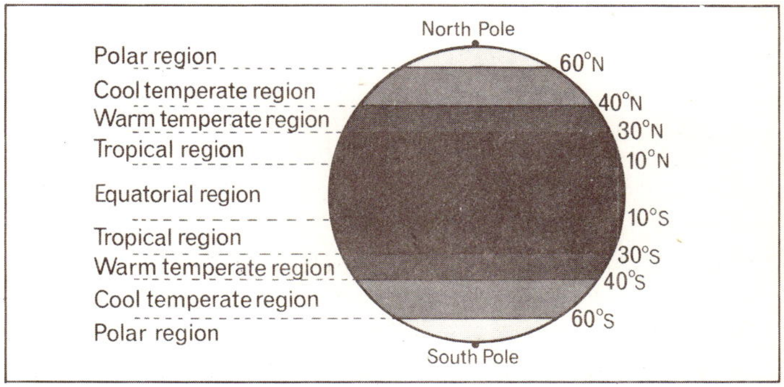

Fig. 72 World temperature regions

STUDY SECTION ON CHAPTER 5

1. Explain briefly the meaning of each of the following terms: insolation, radiation, convection, conduction.
2. Draw diagrams (where appropriate) and write a few words explaining why:
   (*a*) The tops of hills are often covered with snow while nearby lowlands have no snow.
   (*b*) South-facing slopes in Britain are warmer than north-facing slopes.
   (*c*) Places near the equator are warmer than places far from the equator.
   (*d*) Summer is warmer than winter.
3. What is a maximum-and-minimum thermometer?
4. What is a Stevenson screen?
5. What is an isotherm?
6. What is meant by (*a*) annual temperature range
   (*b*) diurnal temperature range.
7. Study temperature maps and statistics to find places with:
   (*a*) a small annual temperature range in tropical latitudes,
   (*b*) a small annual temperature range in temperate latitudes,
   (*c*) a large annual temperature range.
8. The mean annual temperature at the top of Ben Nevis is –0·3°C and at the foot of the mountain it is 8·3°C. Describe fully how these temperatures would be obtained and explain why they are different.
9. The mean January temperature at Valencia (SW Ireland) is 6·7°C and at Moscow it is –10·6°C.
   (*a*) Explain why Valencia is so much warmer than Moscow.
   (*b*) State with reasons how you would expect their temperatures to compare in July. (Check your answer from a table of climate statistics.)
10. Explain with the aid of a diagram why in January the temperature at Athens is 8·9°C while at Leningrad it is –7·8°C.

11 (*a*) Draw temperature and rainfall graphs for San Francisco, Sydney, Singapore, and London.

(*b*) For each of these towns state:

(i) average annual rainfall,

(ii) seasonal distribution of rain (summer, winter, all year),

(iii) maximum and minimum temperatures,

(iv) annual temperature range.

Recording and Forecasting the Weather

6

In the previous chapter the various components which go to make up the weather have been considered individually. It is only when these are all measured at the same time at a weather station that a complete picture of the state of the atmosphere is obtained.

Weather Stations

Weather stations are well distributed throughout the British Isles, many of them being at airports and R.A.F. stations, such as Heathrow (London), Culdrose (Cornwall), Prestwick (Ayrshire), and Stornoway (Hebrides).

At all the stations the weather is recorded on similar instruments operated under well-defined conditions, so that the recordings can be easily fitted together to give a picture of the weather all over Britain.

The main instruments are:

1 Stevenson screen with maximum and minimum thermometers and wet-and-dry-bulb thermometers
2 grass minimum thermometer
3 barometer and barograph
4 rain gauge
5 wind vane and anemometer
6 sunshine recorder

All these instruments except the sunshine recorder have already been described. The **sunshine recorder** is placed in an open space so that the sun's rays strike it from any direction. Every time the sun shines, its rays are concentrated by a lens in the recorder to trace a line on special paper. The length of the trace gives the number of hours of sunshine.

The recordings made by these instruments are entered in a log book at fixed time intervals. At some stations it is every three hours, but if there are only facilities for two readings per day they are made at 9 am and 9 pm.

The information obtained from the instruments is noted in the log book as follows:

barometric pressure	
barometric tendency—rise or fall since last reading	
dry-bulb temperature (air temperature)	relative humidity
wet-bulb temperature	
direction and force of wind	
height, amount, and type of cloud	
visibility	
rainfall	
maximum temperature	These are only recorded at 9 am.
minimum temperature	
grass minimum temperature	

A *Meteorological Register* published by Her Majesty's Stationery Office gives instructions for setting up and reading the instruments.

In order to obtain information about the conditions in the upper atmosphere and to measure the height of the clouds, **weather balloons** (fig. 73) are sent up. These balloons contain a set of instruments for measuring temperature, pressure, and humidity,

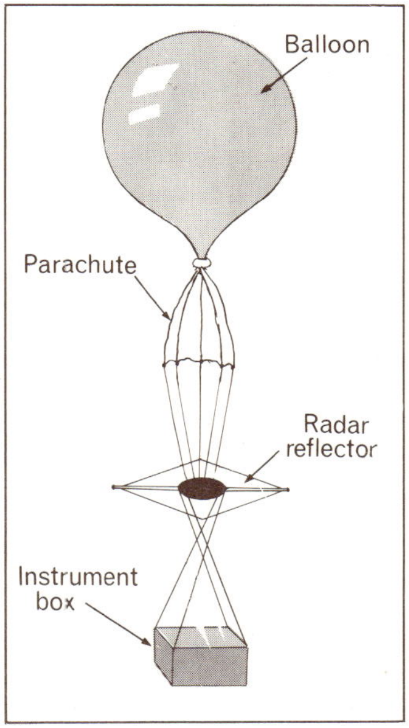

Fig. 73 Weather balloon

and a small radio transmitter for sending back the information to the weather station. They are also fitted with a radar reflector, by means of which the course of the balloon can be followed. This is particularly useful for measuring speed and direction of high-level winds. The balloon is designed to burst at a predetermined height, and a small parachute returns the instrument box to the ground.

The information recorded at the various weather stations is sent in code to the Meteorological Office, where it is used to prepare weather forecasts and to compile climate statistics.

Weather Forecasts

In order to forecast future weather, the meteorologist must first of all have as complete a picture as possible of the present weather over a fairly large area, which often extends beyond national boundaries. The information for building up this picture is obtained from recording stations in many countries, all of which use the same methods of observation and transmit the results in an international code.

One of the main problems in building up a weather picture for western Europe is that there are no islands on which weather stations can be sited in the part of the Atlantic lying west of Europe. Some information is obtained from commercial ships and aircraft, but they are constantly changing position and follow very narrow paths across the sea. In order to fill in some of the gaps **ocean weather stations** have been established. These are operated by specially equipped ships which stay at the positions shown in fig. 74. Britain along with France and Holland is

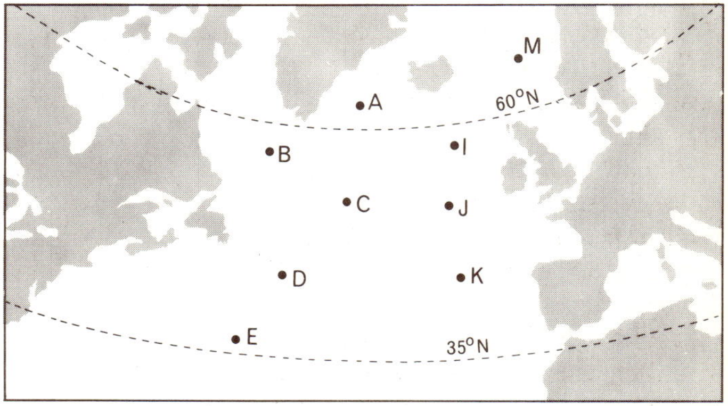

Fig. 74 Positions of the ocean weather stations

responsible for maintaining stations A, I, J, and K. In addition to these ships, specially equipped aircraft fly about 1100 km (700 miles) west from the British coast. On the outward flight they observe the lower air and on the return flight they keep at about 5500 metres (18,000 feet) to take observations in the upper air. During the hurricane season similar aircraft fly from the American coast to track these severe storms. A recent development has been the use of satellites which send back to earth a picture of a very large area of the atmosphere. In future years this method will be increasingly used.

Once all the information has been collected at the forecasting office, it is plotted on a **weather map** using special symbols, some of which are shown in fig. 75. Temperature and pressure are shown by

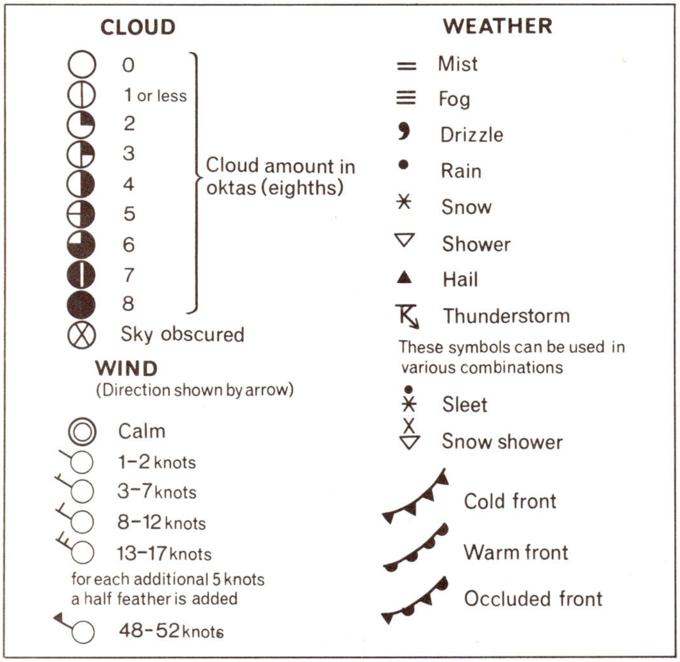

Fig. 75 Some of the main symbols used on weather maps

numbers (degrees and millibars respectively) and all the symbols and numbers are arranged in a standard order as shown in fig. 76. The maps to be used by the weather forecaster are called **synoptic charts** (fig.

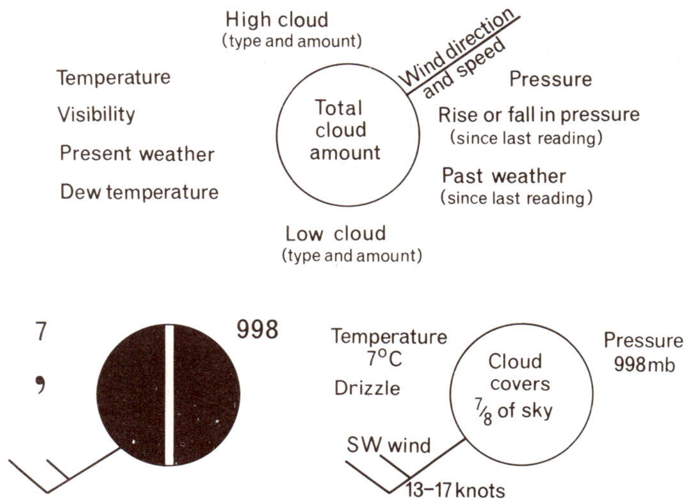

Fig. 76 How information is plotted on a weather map
Note. Visibility is indicated in nautical miles (1 nautical mile = 1853 m = 6080 feet). Dew point is the temperature at which dew will form.
A simplified form of plotting information as used on the Daily Weather Report maps.

77) and have every detail on them, but a simplified form of map is also produced and is available to schools as the **daily weather report** (see Appendix 1). Once all the information has been plotted, isobars are drawn at 4-mb intervals, centres of high and low pressure labelled, and fronts inserted. Speed is essential in this work. With the aid of computers it takes less than an hour for the information to be received, plotted, and the map completed. Detailed maps are produced every hour, so that the forecaster can see how the weather pattern is changing. On the basis of these changes he can forecast what is going to happen in the next few hours: for example, the speed and direction of a depression approaching Britain from the Atlantic can be seen, and its future movements forecast. If a depression suddenly changes its speed or direction just after one of the regular

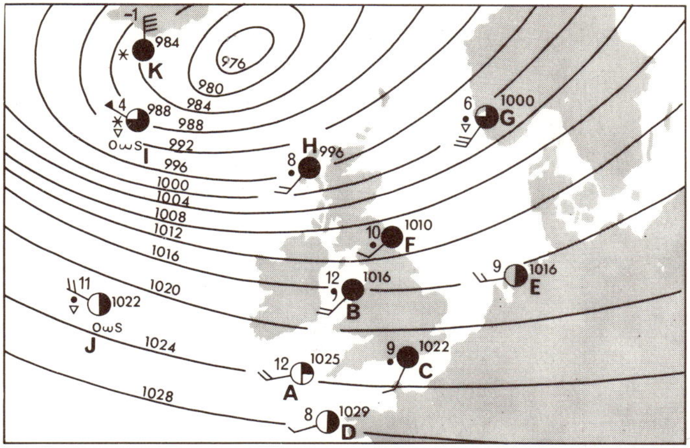

Fig. 77 Simplified version of a weather map Information plotted at each observing station

Station	*Temp.,* °C	*Weather*	*Cloud*	*Wind* knots	*Pressure* mb
A	12	dry	3/8	W 18–22	1025
B	12	drizzle	8/8	SW 18–22	1016
C	9	rain	8/8	S 8–12	1022
D	8	dry	4/8	W 3–7	1029
E	9	dry	5/8	W 13–17	1016
F	10	rain	8/8	SW 8–12	1010
G	6	rain shower	6/8	SW 28–32	1000
H	8	rain	8/8	SW 18–22	996
I	4	snow shower	6/8	NW 48–52	988
J	11	rain shower	4/8	NW 18–22	1022
K	−1	snow	8/8	N 38–42	984

B.B.C. forecasts has been broadcast, an area may have very different weather from what was expected. It is often useful to check on the latest forecast by telephoning the local weather centre (its number will be given under **Weather** in the local telephone directory).

STUDY SECTION ON CHAPTER 6

1 List the main instruments used for measuring and recording the weather.

2 How is information obtained about the state of the upper atmosphere?

3 What are ocean weather stations and why are they necessary?

4 Why does a weather forecaster require a series of weather maps?

5 Use the symbols shown on fig. 76 to plot the following information on a traced outline of the map:

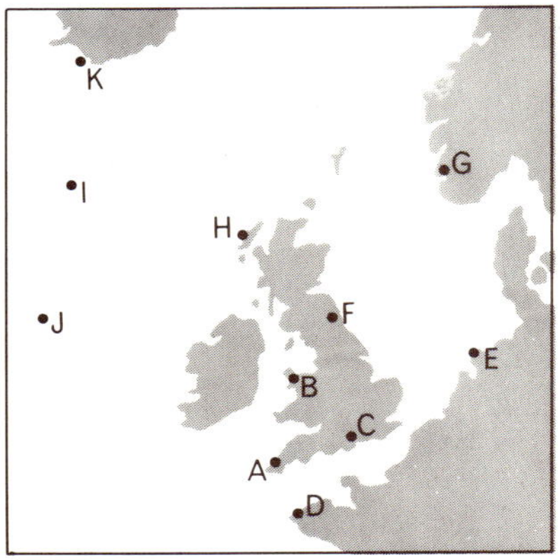

CLIMATIC FACTORS

Station	*Temperature* °C	*Weather*	*Cloud*	*Wind* knots		*Pressure* mb
A	7	dry	0	SE	1–2	1042
B	3	dry	0	calm		1043
C	2	dry	4/8	NE	1–2	1041
D	5	dry	2/8	NE	1–2	1039
E	4	mist	6/8	E	3–7	1040
F	2	fog	6/8	S	1–2	1042
G	3	rain	8/8	SW	33–37	1038
H	8	rain	8/8	SW	18–22	1034
I	10	rain shower	7/8	SW	28–32	1012
J	12	drizzle	8/8	SW	23–27	1026
K	−1	snow shower	3/8	NW	8–12	1004

6 On the map plotted for question 5 draw isobars at 4-mb intervals and shade lightly the high-pressure centre.

7 If the high-pressure area is moving south-east, what would your weather forecast be for station A?

CLIMATE OF THE BRITISH ISLES

part two

General Features

7

Air Masses

Air masses are large bodies of air with distinct physical properties of temperature and humidity. Britain is affected by masses which originate (*a*) in the tropics, and (*b*) in the arctic. Each brings its own characteristic temperature with warm air from the tropics and cold air from the arctic, but the humidity is influenced by the route the air mass follows. If it approaches from the Atlantic (maritime air) its humidity is high, while air coming from Europe (continental air) has a much lower humidity. The principal characteristics of each of these air masses are shown in fig. 78. When one of these air masses is affecting Britain, it usually gives several days (and sometimes weeks) of the same type of weather. Its presence is largely governed by the distribution of the major centres of high and low pressure over the North Atlantic and Europe. Occasionally the north of Britain is affected by a different air mass from the south, and as a result there can be sharp contrasts in the weather between Scotland and England. Figs. 79 a–e show the influence of some common pressure distributions on the air masses. It will be seen from these that even in a narrow country like Britain there can be noticeable differences between east and west. These differences are partly caused by the semi-continuous mountain belt extending from the

Polar maritime air	
Summer	Winter
Cool Showery	Cold Sleet or snow showers

Polar continental air	
Summer	Winter
Warm Dry Hazy in east	Very cold Snow, heaviest in east

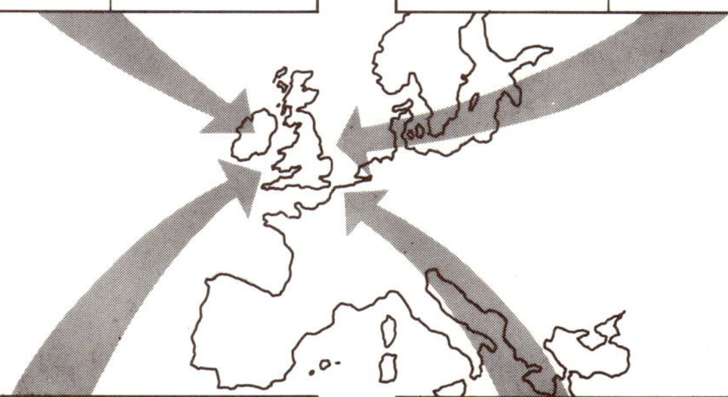

Tropical maritime air	
Summer	Winter
Warm Low cloud Rain or drizzle	Mild Low cloud Rain or drizzle

Tropical continental air	
Summer	Winter
Hot Dry Hazy in east	Mild Dry

Fig. 78 Characteristics of the principal air masses which affect Britain

Scottish Highlands, through the Southern Uplands, Pennines and Welsh mountains, to south-west England. These uplands form a barrier or windbreak between the east and west coasts. As a result, the characteristics of the air masses are slightly modified

Fig. 79a Polar maritime air

This air starts as cold dry air in the Arctic. As it moves over the sea to warmer latitudes it becomes humid and unstable, giving broken cumulus clouds with heavy showers and bright intervals. The north-west winds are usually fresh, becoming strong in the showers. The west coast and hills have most showers, while the east coast is drier and sunnier with moderate winds.

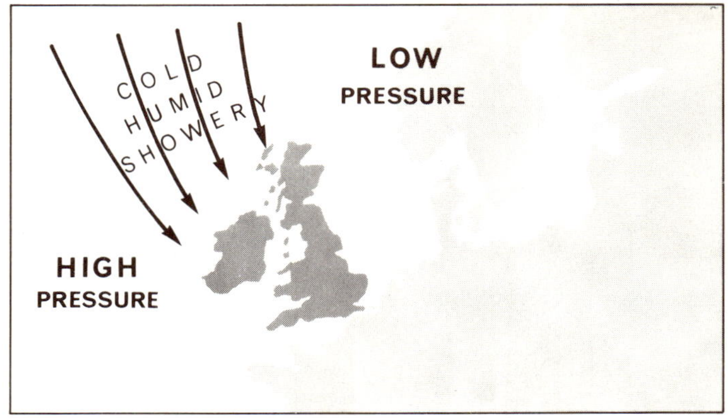

Fig. 79b Polar continental air

The cold dry air from northern Russia is stable. It gives little rain except along the east coast where moisture, collected over the North Sea, is sometimes precipitated as a low layer of stratus cloud or mist. In summer it is warm air when it reaches Britain, but in winter it is very cold as a result of the temperature extremes over eastern Europe. Snow may be brought by this air mass in winter. Western Britain is then much less cloudy and is sunnier than the east.

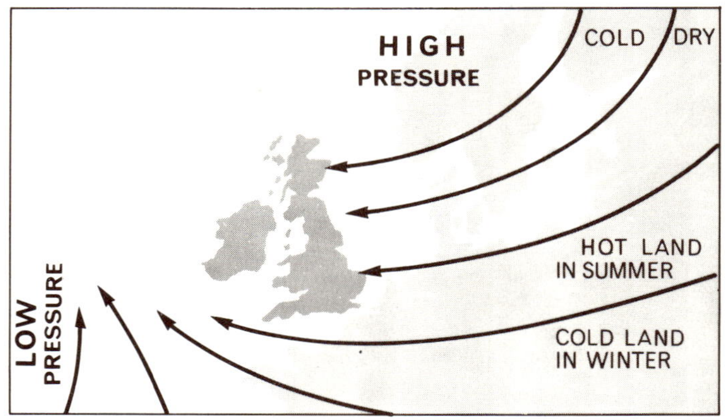

Fig. 79c Tropical maritime air

This warm damp air brings low cloud and muggy conditions to most of the country. Sometimes parts at the extreme east have broken cloud and a little more sunshine. In hilly areas and along the west coast drizzle and sea mist are common.

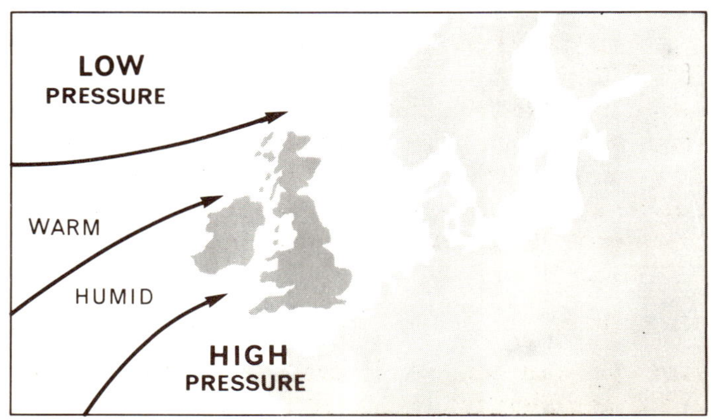

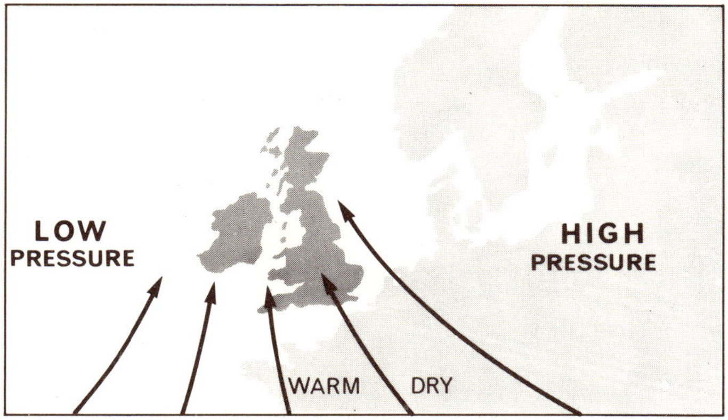

Fig. 79d Tropical continental air
This air often originates over North Africa and brings dry warm weather. Visibility is often poor, especially near the North Sea coast. Scotland seldom receives the full effect of this air.

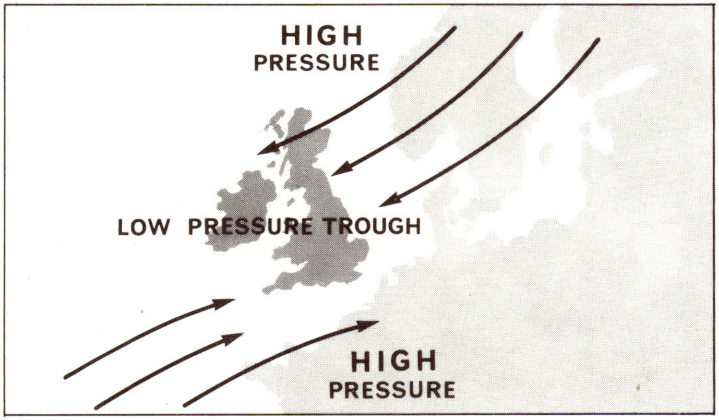

Fig. 79e Two air masses affecting Britain
While southern England is experiencing mild damp weather, Scotland is very much colder, but drier and sunnier.

by the time they reach the leeward side of the country (fig. 80).

Air masses are directed to Britain by the movement of high- and low-pressure centres which often lie hundreds of miles beyond our shores. Sometimes one of these pressure systems is situated near Britain, bringing its distinctive type of weather with it.

Anticyclones

Anticyclones with their gentle pressure gradient and descending air tend to give calm cloudless conditions. In summer it may be very warm, but in winter loss of heat from the ground by radiation can cause a pronounced temperature inversion (fig. 81) with severe frost. If the lower air is humid as a result of passing over the sea or contact with damp ground, radiation fog may form. Although the sun may be able to raise the temperature high enough for the fog to evaporate by mid-morning in spring or autumn, during winter this fog may persist for several days. Sometimes the inversion is over 300 metres (1000 feet) above the ground, and when this happens a layer of fog (or stratus cloud) may form

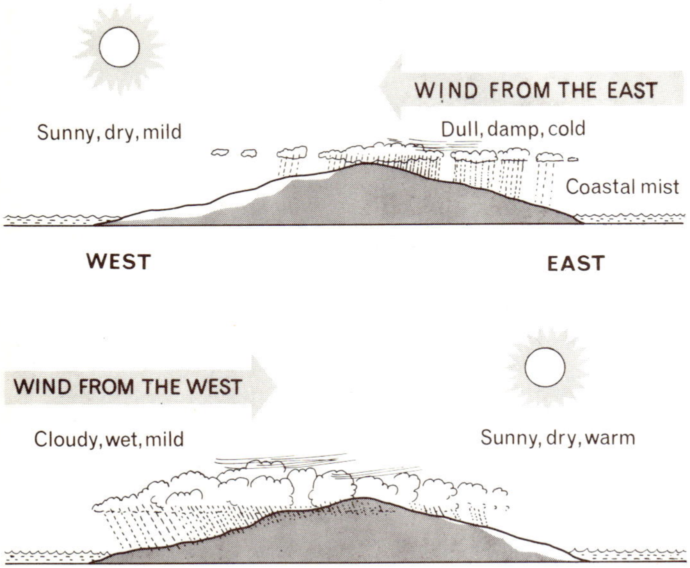

Fig. 80 Differences in weather between east and west coasts

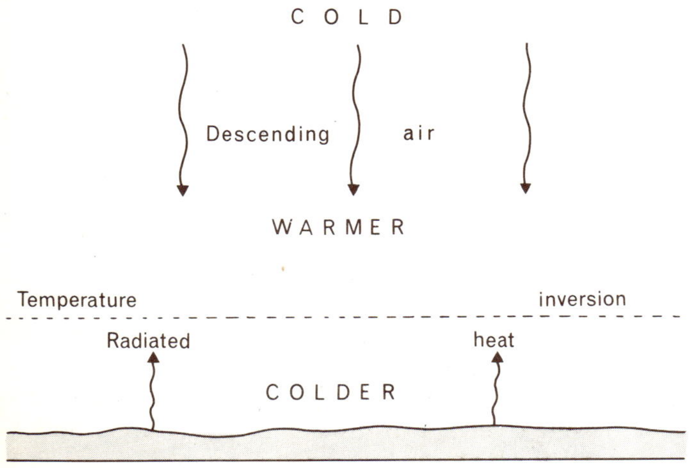

Fig. 81 Temperature inversion during a winter anticyclone

some distance above the ground to give very dull conditions called **anticyclonic gloom**. When skies are clear there often is a large diurnal temperature range, but in persistent fog (or gloom) it remains uniformly cold day and night.

Depressions

In fig. 79e the low-pressure area over northern England is clearly seen to be a boundary between two contrasting air masses. When this situation arises, a **front** is formed between the two bodies of air. The front shown in this figure is the one known as the **polar front**. It is always situated close to Britain, so that the frontal weather associated with this low-pressure region of depressions is one of the most common types we have. The depressions are normally much more extensive than the trough shown on fig. 79e and may cover more than half the country.

The polar front which separates polar and tropical air extends right round the world, and our depressions normally start as a wave in this front many hundreds of miles to the west of Britain. When this wave forms, pressure drops rapidly, especially at the top of the wave, and it develops a system of closed isobars and strong winds. This wave moves eastwards along the polar front. Its leading edge, known as the **warm front**, brings warm air, while its trailing edge is called the **cold front** and brings cold air (fig. 82). Since the cold air is denser than the warm air, it pushes under the warm air. It causes strong up-currents in the unstable conditions associated with the low pressure, and there is considerable precipitation. Fig. 83 shows the weather conditions associated with a depression. The sequence of weather caused by the passage of a depression over Britain is as follows:

1 High feathery cirrus clouds appear in the sky and the wind backs to south-westerly.
2 Cloud thickens and becomes lower until the sun

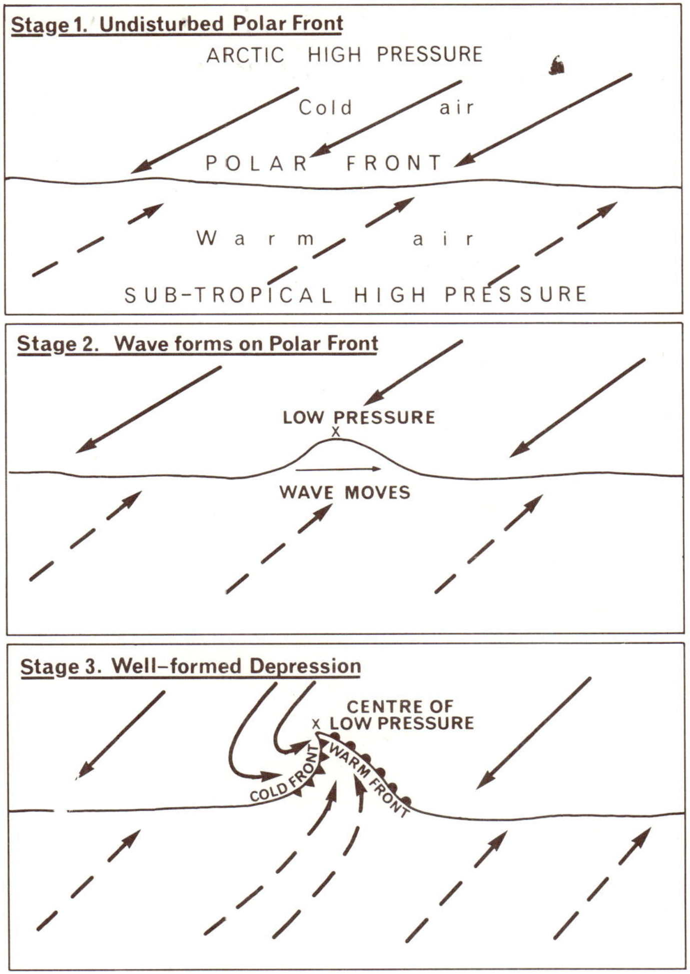

Fig. 82 Stages in the formation of a depression

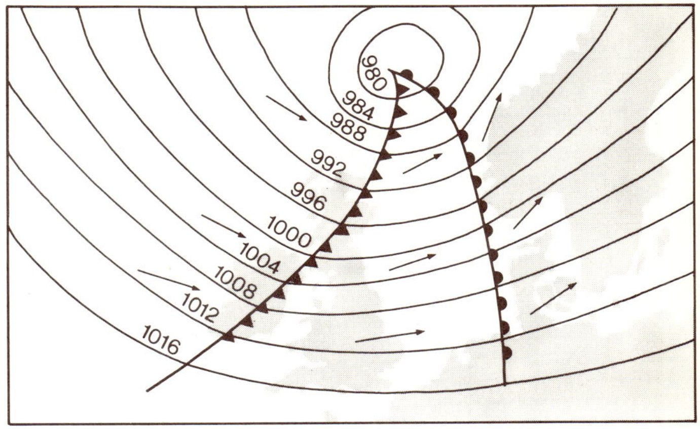

Fig 83a Isobars and winds of a typical depression

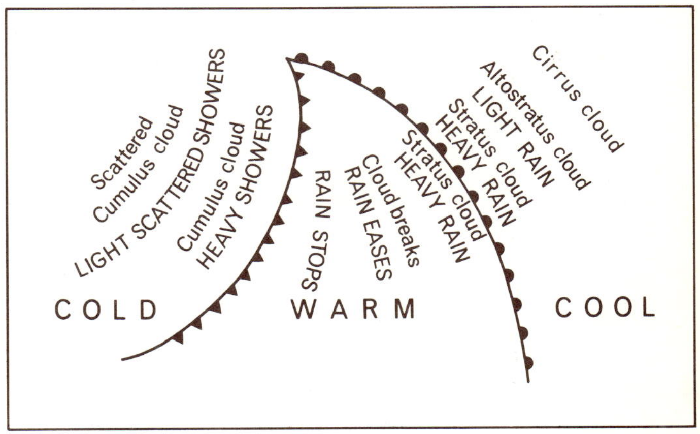

Fig. 83b Weather of a typical depression

is completely obscured. The wind continues to back and freshen. The barometric pressure falls.

3 Low cloud gives steady rain. When the warm front passes, temperature rises and the pressure ceases to fall.

4 Clouds begin to break and rain eases. Wind starts to veer and pressure starts to rise.

5 As the cold front approaches, there is heavy rain from cumulus cloud.

6 The passage of the cold front is accompanied by a sharp veer in the wind to north-west and a marked drop in temperature.

7 Behind the cold front, pressure continues to rise. There are heavy showers, separated by

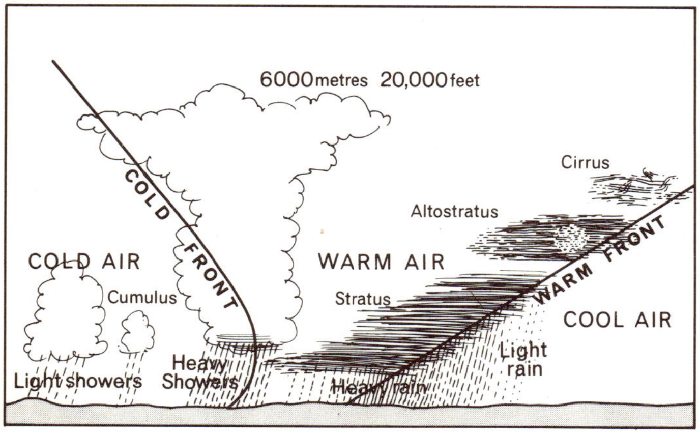

Fig. 83c Cross-section of a typical depression

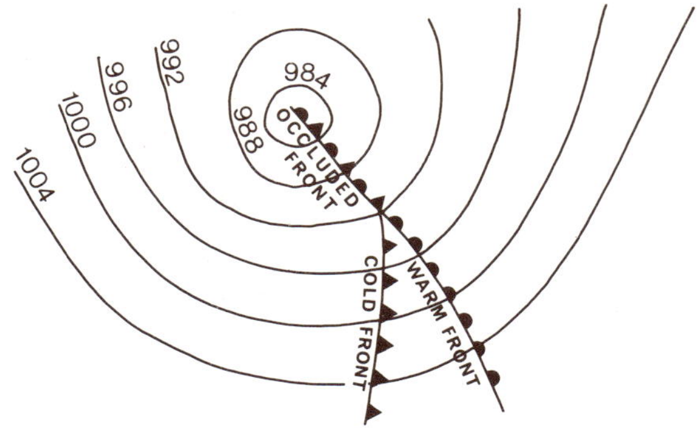

Fig. 84a Arrangement of fronts (at ground level) in a typical occlusion

bright intervals which increase in length as the depression passes away.

The fronts move slowly in an anticlockwise direction round the centre of a depression, with the cold front slowly overtaking the warm front and eventually catching up on it. This gives rise to an **occlusion** (fig. 84) in which the warm air is gradually raised higher into the atmosphere, until the depression ceases to exist.

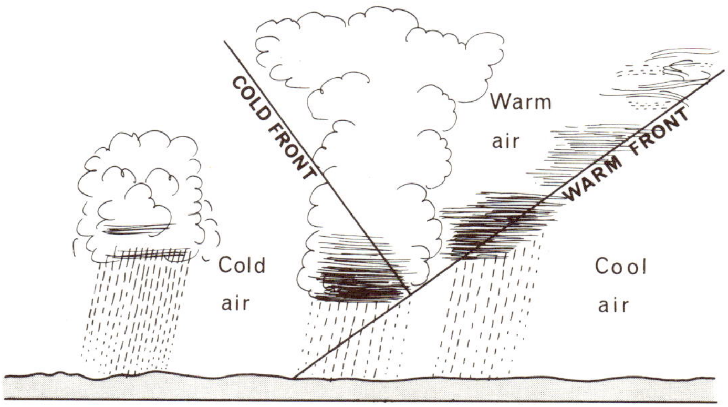

Fig. 84b Cross-section of an occlusion

STUDY EXERCISES FOR CHAPTER 7

1 List the four main air masses which affect Britain's climate.

2 Which air mass produces:
   (*a*) mild cloudy weather in winter,
   (*b*) very cold weather with snow showers in winter,
   (*c*) hot dry weather in summer?

3 Why do anticyclones cause cold weather in winter and hot weather in summer?

4 Fit the following numbers and letters into the correct pairs:

1 The polar front is the boundary between
2 The approach of a depression is indicated by the appearance of
3 The warm front brings
4 The showers behind a cold front come from
5 In an occlusion the .... meet.

A streamers of feathery cirrus cloud
B cumulus clouds
C warm front and cold front
D polar and tropical air
E rain and mild temperatures

Climatic Factors

8

Temperature

The outstanding feature of the temperature of the United Kingdom as a whole is its mildness in winter. This is clearly shown in the table, where the temperatures of coastal places at approximately the same latitude are compared (fig. 85).

	Lat. 54°N	Glasgow	Cartwright	Prince Rupert	Petropavlovsk
JAN.	Max.	6·0°C	−10°C	4·0°C	−5·0°C
	Min.	1·0°C	−19°C	−1·0°C	−11·7°C
JULY	Max.	19·0°C	19·0°C	16·7°C	13°C
	Min.	11·0°C	7·2°C	9·5°C	8·0°C

Glasgow has milder average temperatures than any of the other places for both average maximum and average minimum temperatures. This mildness is caused largely because the most common air movement is from the Atlantic Ocean, and the North Atlantic Drift makes this ocean even milder than it would otherwise be. A similar set of circumstances makes Prince Rupert also mild, but their effect is less marked than in the case of Britain. In summer the influence of the Drift is not so noticeable, but Glasgow is still the warmest of the four places.

The mild winters are of great importance to the people of Britain, and they have played an important part in the development of these islands. In contrast to the 50 million people living in Britain, the Labrador coast is almost uninhabited; and Tierra del Fuego with its black snow-capped mountains and glaciers sweeping down to the coast is nearer the equator than is Scotland. The mildness of the winter weather is of obvious benefit to farmers, as ploughing is often possible in January, and in many areas cattle can be kept in the fields throughout the year. Town dwellers are usually less conscious of weather conditions than country people, but they also benefit from the mild winters. Double glazing and central heating while adding to comfort are not essential, and it is only for short spells that communications are seriously hindered by heavy snowfall.

Within the British Isles there are, of course, many local temperature variations caused by such factors as latitude, altitude, and exposure. Fig. 86 shows temperature graphs for five widely distributed towns, and when they are compared it will be seen that:

1 The total amount of warm weather (average temperature over 10°C) is considerably less in Aberdeen and Stornoway than in the more southerly towns.

2 Plymouth is the only place where the growing season lasts the whole year. Although Stornoway is only 1·5°C cooler than Plymouth in January, it is 2·9°C cooler in August.

3 The amount of cold weather and the temperature range are noticeably greater on the east

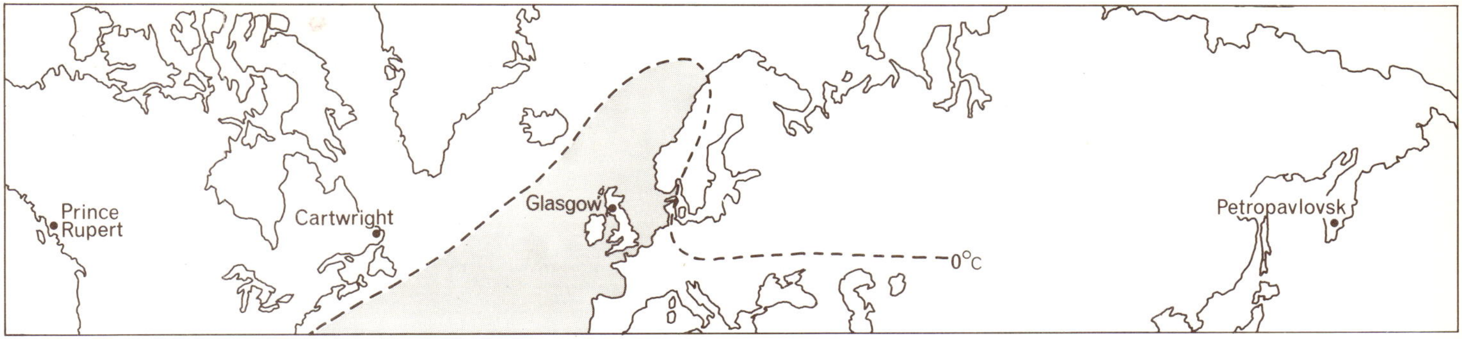

Fig. 85 Location of towns for which temperatures are compared and gulf of winter warmth

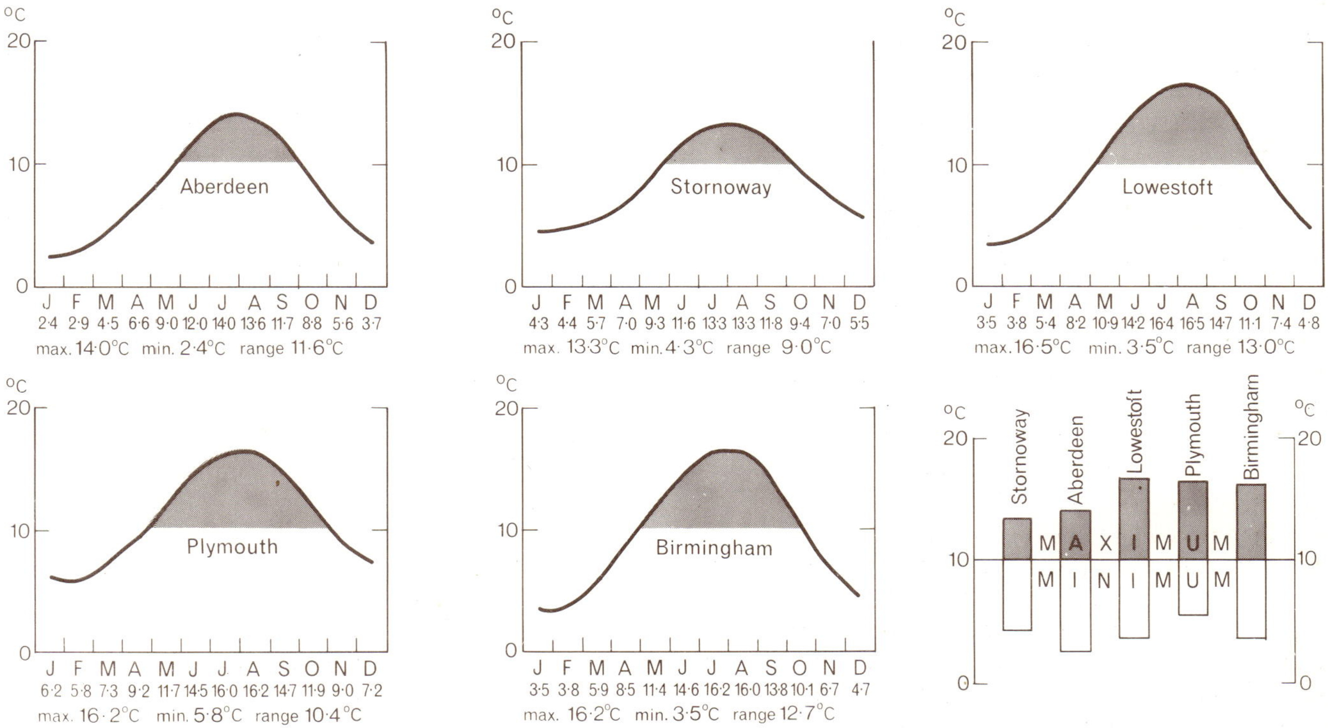

Fig. 86 Temperature graphs

coast (Lowestoft and Aberdeen) than on the west coast (Plymouth and Stornoway).

4 An inland town (Birmingham) has very similar temperatures to those of an east-coast town at similar latitude.

The main reasons for these differences are:

(*a*) Northern places have a shorter summer since they are further from the equator.

(*b*) East-coast places are more extreme, since they are more exposed to continental air which is cold in winter and hot in summer.

(*c*) West-coast places are more equable (less extreme) since they are more exposed to mild Atlantic air.

(*d*) Mountains and uplands provide a semi-continuous barrier which restricts the full effects of Atlantic and continental air to the west and east coasts respectively.

The same temperature pattern is shown on the isotherm maps (fig. 87). In winter, temperatures decrease from west to east, with a slight rise along the east coast due to the presence of the North Sea. In summer, temperatures decrease from south to north, with a slight rise inland due to the greater distance from the sea. (Note that the isotherm maps are based on temperatures reduced to sea-level, while the graphs are drawn for actual temperatures.)

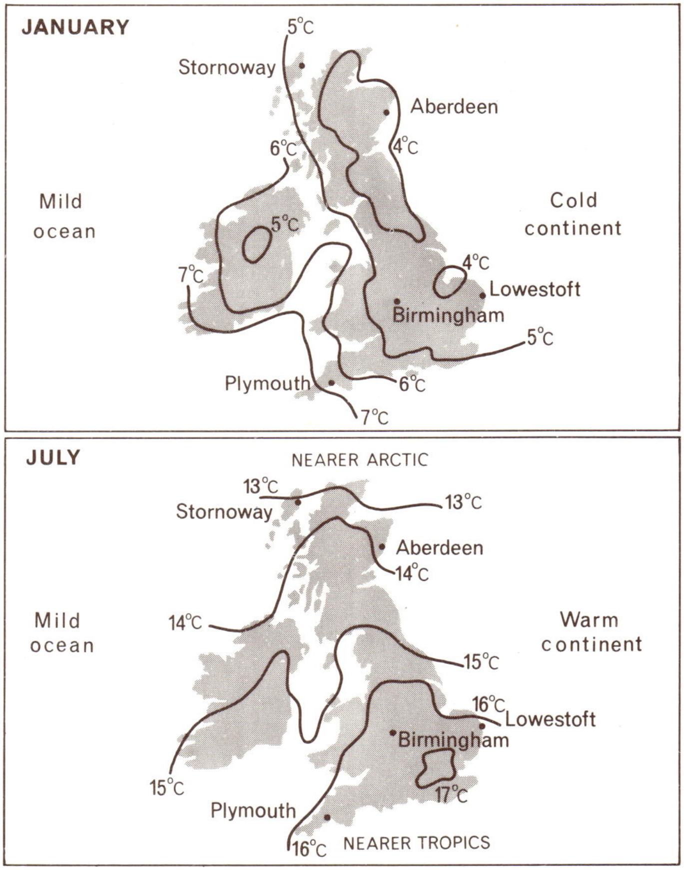

Fig. 87 Isotherms

Rainfall

The moisture which provides Britain's rain is almost entirely of Atlantic origin, and most of the precipitation is caused by depressions crossing the country from west to east. Since many of these depressions are 'filling' and becoming less active when they reach this country, they tend to give less rain in the east than in the west.

Moist air from the sea gives rise to **orographic rainfall** when it crosses hills. As a result all uplands are noticeably wetter than adjacent lowlands. Where uplands are extensive, as in Scotland, they considerably reduce the humidity of the air and there is a marked rain shadow on the leeward side of hills. This rain shadow is particularly well marked on the east side of Britain, since the prevailing winds are from the Atlantic. When the wind is from the east, west-coast areas are often sunny and dry, while the east is cloudy and wet.

Convection rainfall is much less common than the other two types, but it does sometimes occur, especially in southern England, at the end of a hot dry spell in summer. The pressure starts to fall, and the air becomes unstable, leading to heavy convectional downpours and thunderstorms.

These three causes produce a definite pattern of rainfall over the country which is illustrated in fig. 88.

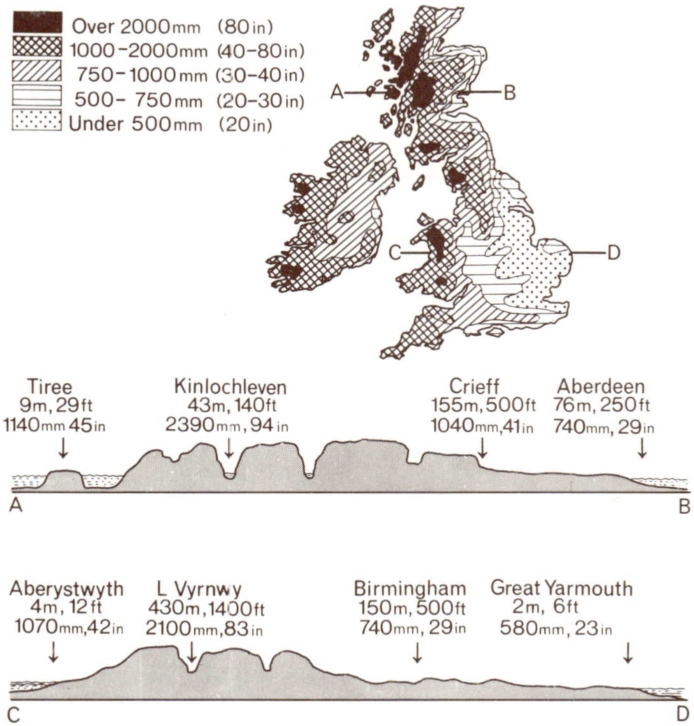

Fig. 88 Average annual rainfall

Birmingham, for example, receives much of its water from reservoirs in the Welsh mountains, and Glasgow's main water supply comes from Loch Katrine in the Highlands. The other principal use made of water in Britain is for the production of hydro-electricity. There are many power stations in the wetter parts of the Scottish Highlands, such as at Loch Sloy near the head of Loch Lomond.

The farmer, the industrialist, the town water-engineer, and the hydro-electricity board are not only concerned with the total annual rainfall but also with how this rain is distributed over the various months of the year. Although Britain has no dry season, there are slight seasonal variations as shown in fig. 89. It is clearly seen that in all parts of the country the spring and early summer months are drier than the rest of the year, with the highest rainfall totals coming in late autumn and winter. The drier weather of spring is often accompanied by cold north-east winds and poor visibility which result from a well-developed high-pressure area over the cold Baltic Sea and Scandinavian mountains (fig. 78). At this time the polar front with its depressions and tropical air lies south of Britain. As the Scandinavian high pressure collapses in late spring, the polar front moves north and the typical broken, rather wet, summer weather arrives. For the rest of the year there is no regular pattern to the weather. The normal variable conditions are brought by a series of depressions, short spells of more settled weather from one or other of the air masses, and occasional anticyclones.

The great variability of weather from year to year and the caution with which average figures should be used are emphasized in the following table:

Recording Station	*Rainfall in August*			*Sunshine in August* (in hours)	
	Average	1947	1948	1947	1948
Oban	118 mm 4·65″	3 mm 0·11″	152 mm 5·98″	262	95
Eskdalemuir (Dumfriesshire)	140 mm 5·50″	1 mm 0·03″	254 mm 10·00″	260	70
St. Andrews	71 mm 2·79″	13 mm 0·51″	128 mm 5·03″	203	105
Nairn	75 mm 2·94″	8 mm 0·30″	109 mm 4·30″	243	67

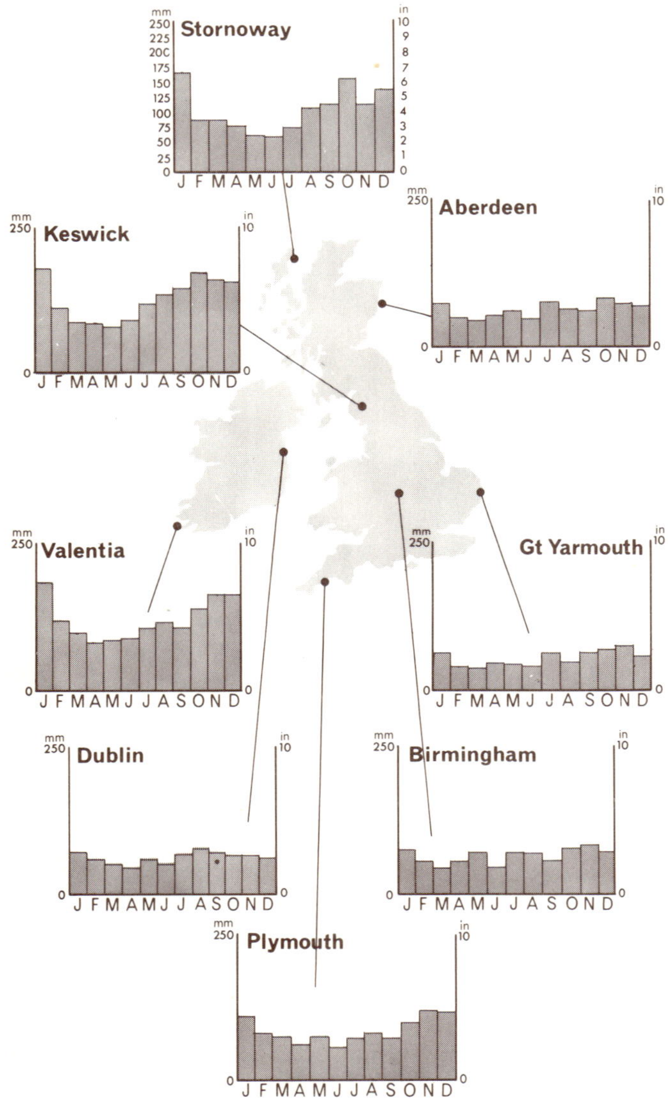

Fig. 89 Pattern of monthly rainfall

It will be seen that the conditions which resulted in the exceptionally dry weather in August 1947 and the exceptionally wet August of 1948 affected the whole of Scotland. The close connection between sunshine and rainfall (cloudiness) should also be noted.

When considering the value of rainfall, the loss by evaporation and percolation must be taken into account. The average loss by evaporation is about 380 mm (15 inches) per year so that, although the evaporation rate will vary greatly with temperature changes over the year, all parts of Britain receive more rain than they lose by evaporation. It is only in south-east England, where the annual rainfall is about 500 mm (20 inches) in places, that there is a danger of water shortage for farming during the hot summer months. In these areas a little irrigation is practised to improve the quality of the crops, and to insure against a dry spell. Elsewhere in Britain the water supply is adequate for all farming requirements. In some of the wetter areas it is too cloudy and wet for large-scale grain growing, although the rainfall gives good pasture for cattle.

Loss of rain by rapid percolation depends on the nature of the rock. It is particularly rapid in the permeable chalk and limestone areas of south-east England. Even when rainfall is heavy, there is a lack of surface water on hills like the Cotswolds and Downs. These areas are often used as grazing land with well-scattered farms and few villages. The edges of the uplands, where the underground water reappears in the form of springs, are often the sites of old villages—located there partly because of the reliable water supply.

In the past, most towns and villages had their own local water supply from springs and wells, but modern cities and large industrial concerns with their need for a large and reliable supply of water can no longer depend on local sources. Fortunately no part of Britain is far from an upland area of heavy rainfall, where the nature of the landscape makes the construction of storage reservoirs relatively simple.

Snow

The amount of snow which falls over Britain is affected by both the humidity and the temperature of the air, so that snow is most likely when cold damp air is blowing from the arctic in winter. This air will not give snow, however, unless conditions for precipitation are present. In hilly areas the uplift caused by the mountains, along with the lower temperature found at high altitudes, results in much heavier and more frequent snowfalls than over lowlands. This might be referred to as *orographic snow*. Widespread snow on low ground is most common when the cold arctic air enters the circulation of a depression. The cooling effect of this air produces *cyclonic snow*. On the higher hills snow can fall almost any time between September and May, while in some places pockets of snow lie all summer in north-facing hollows. On the lowlands snow is uncommon between April and November, and there is a noticeable difference between the milder west and cooler east. There is also a slight increase in snowfall towards the north. These features are illustrated on fig. 90.

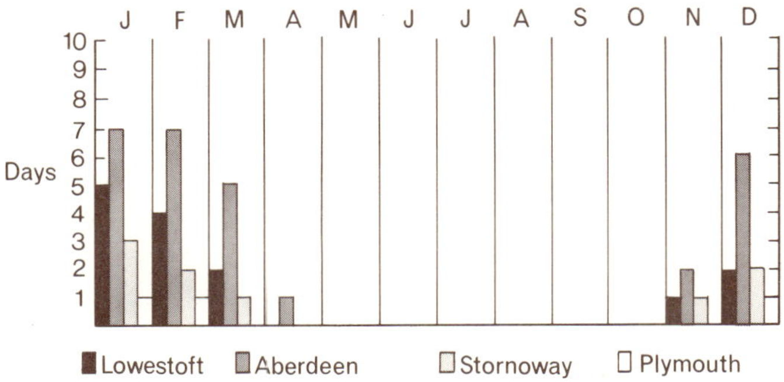

Fig. 90 Average number of days with snow lying

Wind

Like all the other elements of British weather, wind varies greatly in both speed and direction, but once again there is a pattern which appears in the following tables.

Average Wind Direction

N	NE	E	SE	S	SW	W	NW	Calm
8%	6%	5%	4%	20%	15%	18%	9%	15%

Average Wind Speed in Knots

	J	F	M	A	M	J	J	A	S	O	N	D
Lerwick	17	16	16	14	12	11	11	11	13	15	16	17
Birmingham	9	8	8	8	8	7	7	7	8	7	8	9
Plymouth	11	11	12	9	10	8	8	8	10	8	10	11

Average Number of Days per Year with Gales

Lerwick	Glasgow	Scilly Isles
38	7	25

The frequency with which winds blow from a direction between south and west is mainly due to the large number of depressions crossing Britain. Since in many cases the centre of low pressure lies to the north of the country, it is the southerly part of the anticlockwise air circulation which brings these winds. The statistics of wind speeds and gales show quite clearly that coastal areas have much stronger winds than inland stations, and that there is a tendency for strong winds to be more common in the north than in the south. The pattern of wind speeds over the year, however, is similar in all parts of the country, with the winter months noticeably windier than the summer ones.

STUDY SECTION FOR CHAPTER 8

1 At which season has the North Atlantic Drift its greatest effect on Britain's temperature?

2 Which regions of Britain have:
(*a*) the most extreme temperatures,
(*b*) the mildest winters,
(*c*) the coolest summers?

3 Give two reasons to explain why western Britain is wetter than eastern Britain.

4 What are the three most common wind directions in Britain?

5 In which parts of Britain are gales most common?

6 On an outline of the northern hemisphere print the average January temperature (calculated from the maximum and minimum temperatures on page *55*) at the positions of Prince Rupert, Cartwright, Glasgow and Petropavlovsk. Explain why the Glasgow average is higher than those of the others.

7 Explain why the temperature pattern as shown by the isotherms in fig. 87 is different in January from what it is in July.

8 Show graphically the average, 1947, and 1948 rainfalls for Oban, Eskdalemuir, St. Andrews, Nairn, given on page 57.

Climatic Regions

In a small country like Britain the general pattern of the weather is the same everywhere. The reason for this is that the major pressure distributions in the atmosphere are so large that on most occasions the whole country is under the influence of the same air mass or pressure system. The difference of 10° latitude between Shetland and Land's End, and the semi-continuous mountain barrier from the north of Scotland to the south-west of England do, however, give us noticeable regional differences which make it possible to recognize a number of distinct climatic regions (fig. 91).

Uplands

Uplands, in whatever part of the country they are situated, have a very distinct type of weather which separates them off as different regions from neighbouring lowlands. Temperatures are lowered by altitude, and in most cases the growing season is so short that cultivation is impossible. Clouds and rainfall are very common as, in addition to the general rain from depressions, there is orographic rain, and occasionally convectional rain when the steep slopes trigger off strong up-currents in an unstable atmosphere. The large amounts of cloud greatly reduce the hours of sunshine and, when the cloud base is low, hill fog often occurs. Because of the low temperatures,

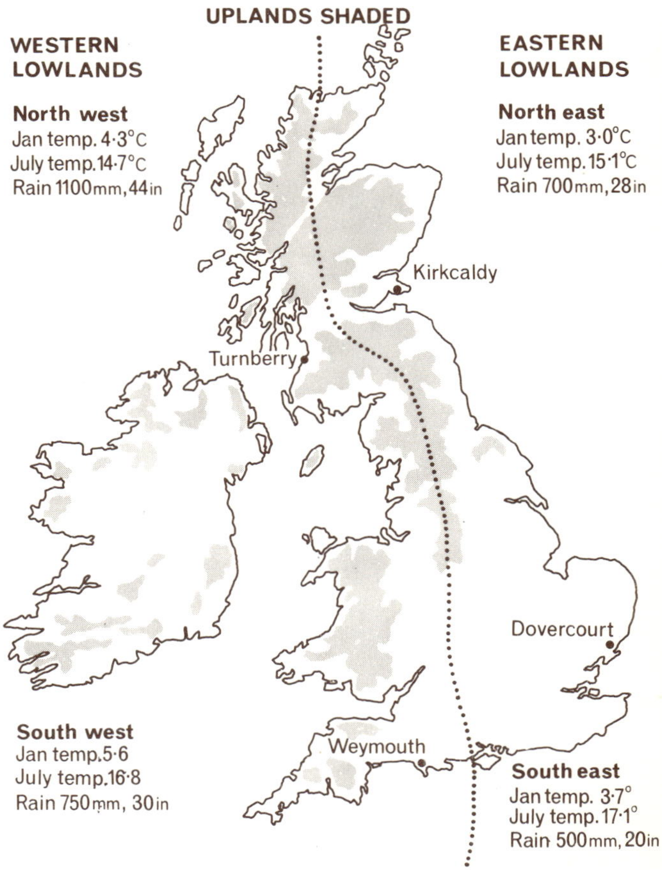

Fig. 91 Climate regions of the British Isles

snow is common, and for much of the winter the rough pastures cannot be used for grazing sheep or cattle. As shown in fig. 66, page 37, there is a relatively small proportion of land at higher altitudes. Since there is less friction and shelter than at lower levels, wind speeds are generally high with very few calm days.

The Scottish Highlands form one of the most extensive upland areas in Britain. While it must be remembered that other factors, such as very hard rock, glaciation, the 'highland clearances' and difficult communications, have influenced the development of the region, climatic factors have also played an important part in producing the present cultural landscape. Farms are mostly concerned with stock rearing, and a major problem is the provision of winter feed. The upland pastures have a poor carrying capacity, even in summer. In winter they are covered with snow for long spells, while in the glens very limited sunshine, as a result of frequent cloud cover and shadow from the mountains, makes the ripening of crops difficult. As a result, farms are less numerous than in most other parts of Britain, and every autumn large numbers of sheep and cattle are sold at markets in bordering towns such as Perth and Oban.

The abundant winter snowfall does have its advantages, however, and the Cairngorms area is Britain's leading winter sports region. Considerable development has taken place around Aviemore where, in addition to chair lifts and ski tows, a modern hotel and chalet accommodation have been provided, and there are facilities for indoor entertainment such as skating and dancing.

At the head of Loch Lomond (fig. 92), where the average annual rainfall is over 2500 mm (100 inches), a large reservoir has been constructed at Loch Sloy 286 m (940 feet) up in the mountains and from it water is fed by pipes to a generating station on the west shore of Loch Lomond. The power is then transmitted by overhead cables to the industrial area around Glasgow.

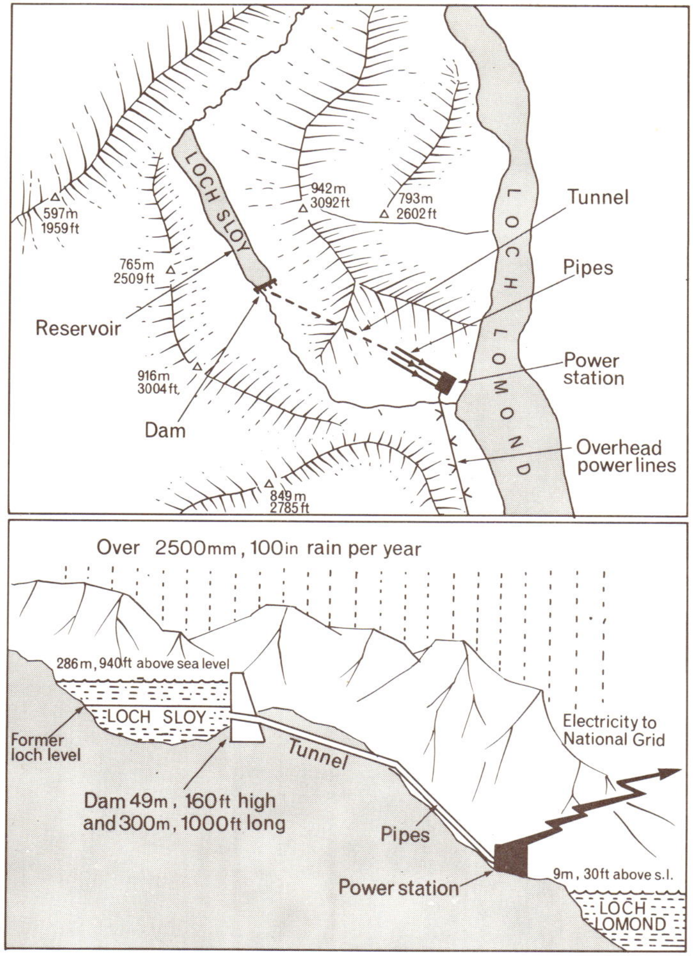

Fig. 92 Loch Sloy hydro-electric power scheme where heavy rainfall has been used along with mountainous topography to produce electricity

Two industries which have been introduced into the highlands because of the heavy rainfall are aluminium processing at Kinlochleven and Fort

William and the woodpulp mill for paper making at Corpach (fig. 93).

The heavy rainfall of mountainous areas often benefits places many miles away. For example, the reservoirs of the Elan valley in central Wales supply

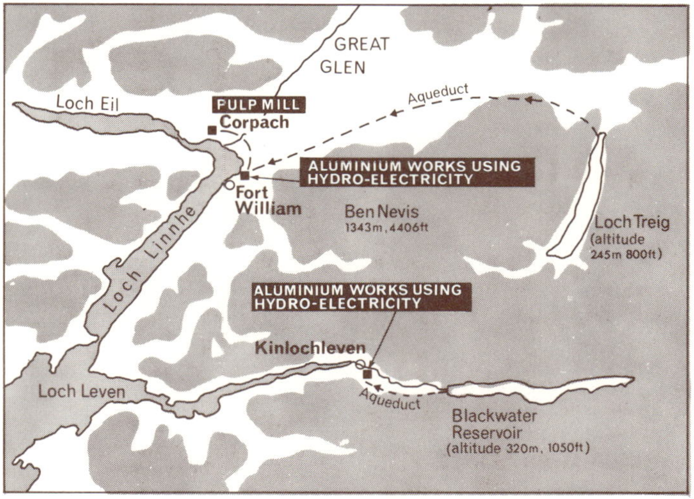

Fig. 93 Three places—Corpach, Fort William and Kinlochleven—where heavy rainfall is used for industry.

Birmingham with much of its water, and Lake Vyrnwy in Montgomeryshire is a source of water for Liverpool.

Western Lowlands

The west side of the country is the first part to be affected by Atlantic air. As a result, the weather tends to be humid with fairly high rainfall and considerable cloud, even when it is not raining. The heavy rainfall is caused by the proximity of large uplands, as well as by the western exposure. This can be seen from the fact that the low-lying island of Tiree has only 1140 mm (45 inches) per year while Fort William at the foot of Ben Nevis has 2000 mm (79 inches). When easterly winds are blowing, the west-coast lowlands have a pronounced rain shadow effect, with much less cloud than normal. This is one of the reasons for the sunny weather commonly experienced in north-west Scotland in spring when easterly winds are fairly common.

The Atlantic air also tends to give temperatures of a small annual range. Once again the mountain barrier limits the effect of easterly winds, and this is of particular importance in winter when they tend to be considerably colder than winds from the west.

This western region of the country can be subdivided into a cooler northern part and warmer southern part. A comparison of the temperature averages for Turnberry (Ayrshire) and Weymouth (Dorset) shows this difference very clearly.

Town	*Latitude*	*January*	*July*	*Year*
Turnberry	55°19′N	4·3°C	14·7°C	9·3°C
Weymouth	50°36′N	5·6°C	16·8°C	11·0°C

Eastern Lowlands

Although eastern Britain, like the rest of the country, has its weather dominated by Atlantic air, it is to the lee of the hills. As a result it is noticeably drier and less cloudy than the west. The east, however, is exposed to the cold winds from the Continent, which lower the winter temperature. This can be seen when the following statistics are compared with those for Turnberry and Weymouth.

Town	*Latitude*	*January*	*July*	*Year*
Kirkcaldy (Fife)	56°08′N	3·0°C	15·1°C	8·2°C
Dovercourt (Essex)	51°57′N	3·7°C	17·1°C	10·2°C

Although summers in the east are a little warmer than in the west, the difference is less marked than in the winter.

July	Kirkcaldy is 0·4°C warmer than Turnberry
	Dovercourt is 0·3°C warmer than Weymouth
January	Kirkcaldy is 1·3°C colder than Turnberry
	Dovercourt is 1·9°C colder than Weymouth

These average temperatures also show that, as in the west, the north is colder than the south throughout the year, so once again the country can be subdivided into a northern and a southern region.

The places taken as examples of the four lowland climatic regions are all coastal, but in a narrow country such as Britain there is not sufficient difference between coastal and inland parts to justify further regional division.

Although the uplands provide grazing land for sheep and some cattle, most of Britain's farming takes place on the lowlands where conditions of relief, soil, and climate are much more favourable. The map of fig. 94 shows that on these lowlands there are marked contrasts between one part of Britain and another. Although differences in soil and accessibility provide two of the reasons for the differences, the change in climate from one region to another is the most important reason.

Over much of the west and south-west, grass—both permanent and rotation—is the most common field crop, and cattle are very numerous. Crops are also grown in rotation with the grass, but they are mostly for fodder; wheat is uncommon. The abundant rainfall, limited sunshine, and mild winters all favour this type of farming.

In contrast, the eastern side of the country is the main arable area, where barley, oats, wheat, and sugar beet are common. Grass also appears in the rotation of eastern farmers. It is seldom left for more than three years without being ploughed, while on the west coast a field may be left in grass for as long as ten years. The difference in crop patterns reflects the lighter rainfall and higher proportion of sunshine experienced in the east, and these dry sunny conditions favour ripening and harvesting. Cattle are also fairly common on east-coast farms, especially dairy herds around large towns. In north-east Scotland

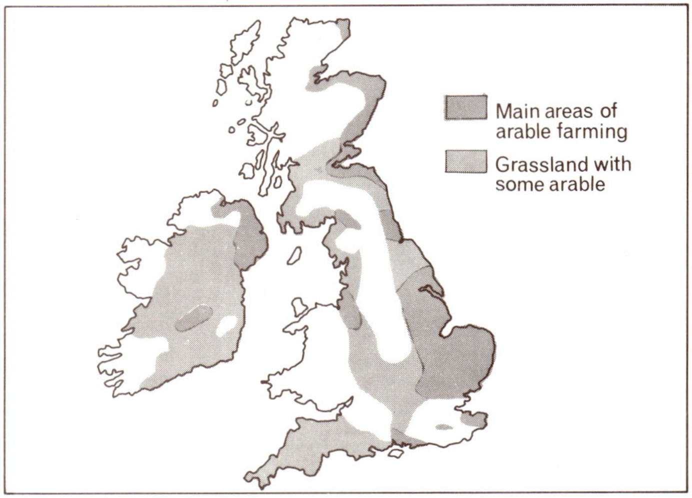

Fig. 94 Simplified map of arable and grassland areas

much of the barley is used for fattening beef cattle, but it is the sale of crops which provides most of the farm income, especially south of the Firth of Tay.

Crops reflect temperature differences much more readily than do animals. While there is little noticeable difference between dairying in Ayrshire and in Somerset, wheat and sugar beet become less common north of the Scottish border where they are replaced by barley and swedes.

Factors other than climate also have a pronounced effect on lowland farming. For example, south of London sheep rearing on the chalk Downs and dairying in the clay vales are partly in response to differences in soil and drainage; while in East Anglia, despite the suitability of the climate for cereals, fodder crops for dairy cattle and market-garden produce are of great importance, due to the profitable London market for milk and fresh vegetables.

STUDY SECTION ON CHAPTER 9

1 What are the two main factors which result in noticeable regional differences in Britain's climate?

2 Name three common characteristics of the climate of each of the following regions:
(*a*) uplands
(*b*) western lowlands
(*c*) eastern lowlands

3 Study the following statistics and suggest whereabouts in Britain you think each station is located.

Station	*Jan. Temp.*	*July Temp.*	*Annual Rainfall*	
A	3·3°C	13·9°C	690 mm	27″
B	3·9°C	17·8°C	580 mm	23″
C	7·8°C	16·1°C	860 mm	34″
D	3·9°C	13·9°C	2030 mm	80″

4 For each station in question 3 describe the climate factors which give it its characteristic features.

10 Local Climate

In this book, as in most geographical texts, the climate of an area is normally described by giving the average temperature and rainfall statistics. While these figures are very useful for seeing the general climate pattern, there are many important local features which appear only when the climate is examined in more detail. The significance of local conditions can be best appreciated by taking one small area and examining the climate in some detail.

The Glasgow Basin

When looking at local climate the physiographic situation of the place must be considered, as it plays an important part in determining the type of weather experienced. Fig. 95 shows that Glasgow lies in a basin surrounded by hills which rise to over 300 metres (1000 feet). Two important gaps in these hills are in the north-east, where a low pass leads to the Forth valley, and in the west where the Clyde flows between the Kilpatrick and Renfrewshire hills.

RAINFALL

The annual pattern of rainfall shown in the monthly averages for Glasgow Airport is repeated throughout the area, with a pronounced minimum from May to July. Fig. 96 shows that the annual total varies noticeably from one part of the basin to another. Most rain is brought by winds from a westerly direction, and the rain shadow effect of the Renfrewshire hills is very pronounced, especially in the eastern part of the area. The surrounding uplands all have averages of over 1250 mm (50 inches) and this has proved most useful in providing an ample water supply for Glasgow and its satellite towns. Ordnance survey maps of the area show numerous reservoirs in these hills.

When studying local rainfall and its effects on the people of an area, it is not just the amount and seasonal distribution which are important. Other factors, such as the intensity and frequency of driving rain, are of considerable significance. A fall of 25 mm (1 inch) may be spread over 12 hours, or it

Average Climate Statistics for Glasgow Airport

	J	F	M	A	M	J	J	A	S	O	N	D	Total
Rainfall (mm)	117	81	64	58	66	61	79	84	91	119	104	107	1031
(inches)	4·6	3·2	2·5	2·3	2·6	2·4	3·1	3·3	3·6	4·7	4·1	4·2	40·6
Temperature °C	3·1	3·9	5·6	7·8	10·7	13·6	15·1	14·8	12·6	9·5	6·1	4·4	8·9

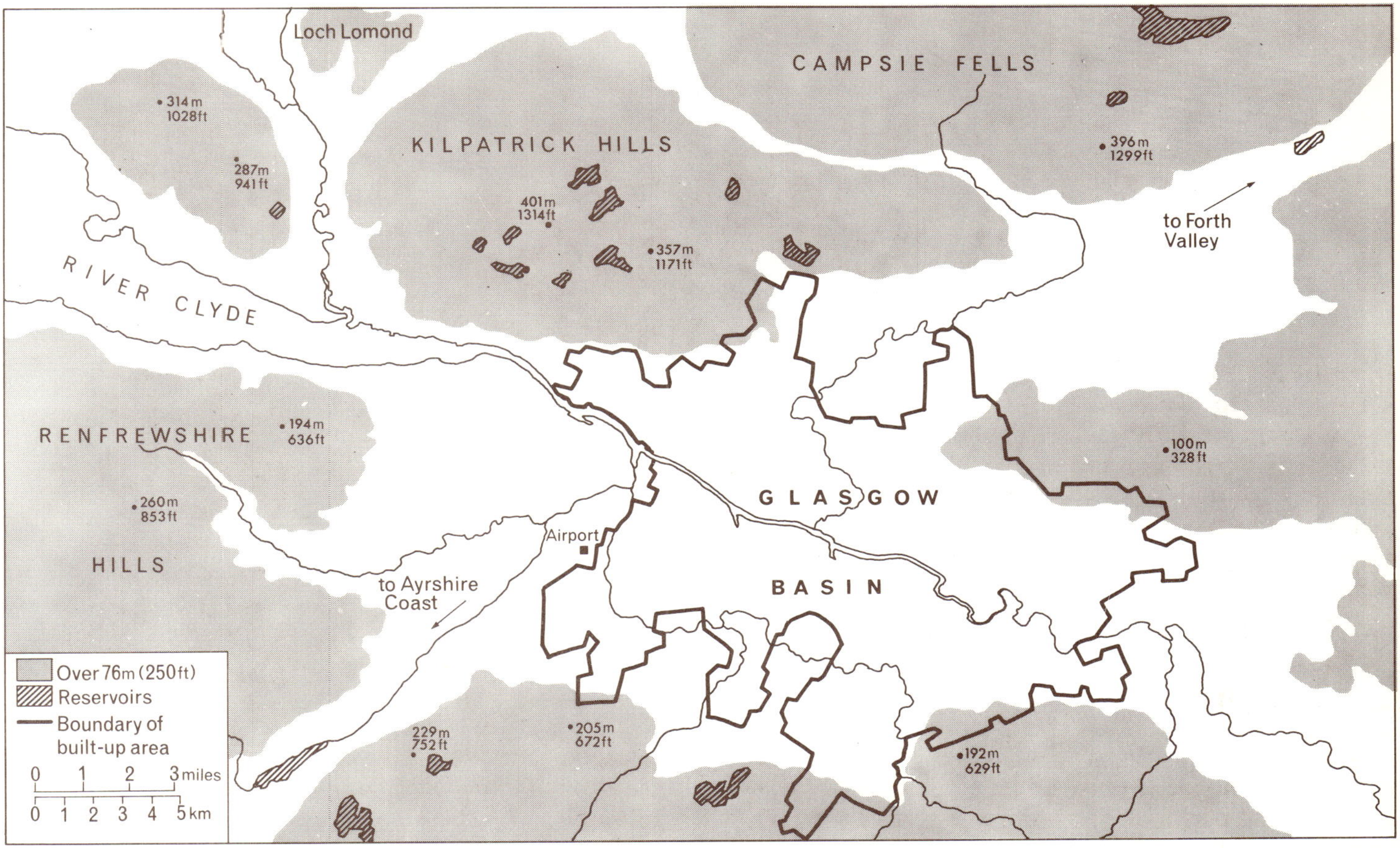

Fig. 95 Physical features of the Glasgow basin

may all be concentrated in one hour The latter can mean damage to standing crops, and the adequacy of the drains in a town will be tested. Rain driven at an angle to the vertical by strong winds is not only much more unpleasant physically; it also penetrates pores and cracks in buildings, causing considerable damage in the course of time. In the Glasgow area, for example, the west- and south-west-facing sides of buildings are more subject to attack from driving rain than other sides. Householders are often very conscious of this when water penetrates badly fitting window frames.

TEMPERATURE

The average temperature statistics show a mild winter and cool summer. When compared with similar averages for other parts of Britain, the general picture is quite accurate, but people living in the Glasgow

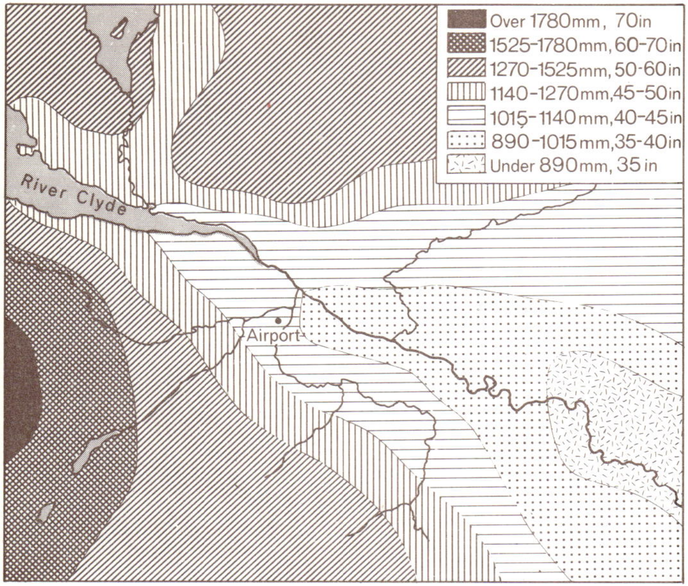

Fig. 96 Rainfall map of the Glasgow basin

area experience variations from these averages which are of some significance.

The following table of extreme temperatures shows some important features, such as the great variations which can be experienced in any month of the year.

Everyone living in the area knows that there can be very cool days in midsummer and spring-like temperatures in midwinter. Midsummer day in June 1972 was colder than midwinter day in December 1971. These short-term variations can be of great significance to the gardener as well as the holiday-maker. Although the average winter temperatures are above freezing, ground frost can occur in any month (though seldom in July and August). Some mild days in February and March may encourage fruit trees to bud but can be followed by disastrous frosts in April. The summer holiday months of July and August can have delightful warm weather one week, followed by cool days when scarves and gloves are essential outdoor wear.

From the viewpoint of damage to crops and dangerous road conditions, the probability of frost is very significant. Although it is not unknown in May, June, September and October, it is so uncommon that for practical purposes it is insignificant. From November to April, however, frost is a significant feature of the climate.

Earliest date of first frost—21st September
Average date of first frost—20th October
Average date of last frost—27th April
Latest date of last frost —19th May

Before we leave this topic we must emphasize the distinction between ground frost and air at or below freezing-point at the level of the thermometers in a Stevenson screen (about 4 feet). The official temperatures all refer to air temperature at four feet. These are not necessarily the conditions on the surface of a road or on the leaves of a small plant. On a clear night there can easily be a temperature inversion at a height of one inch, six inches, one foot, etc., and then the ground temperature will be below the 4-foot temperature. The earliest date of first frost may be 21st September, but this does not mean that there may not be ground frost in August.

Extreme Temperatures at Glasgow Airport

	J	F	M	A	M	J	J	A	S	O	N	D
Absolute highest °C	13·3	13·9	21·1	21·7	26·1	30·0	28·9	29·4	26·7	23·9	16·7	14·4
Average day °C	5·4	6·6	9·0	11·9	15·4	18·0	19·0	18·8	16·4	12·6	8·8	6·6
Average night °C	0·8	1·2	2·3	3·8	6·1	9·1	11·1	10·8	8·8	6·3	3·4	2·2
Absolute lowest °C	−17·8	−15·0	−13·9	−7·2	−5·0	−1·7	1·7	0·6	−3·9	−8·3	−8·9	−11·1

WIND

The importance of wind was brought home to the people of the Glasgow area in January 1968 when a severe storm caused great structural damage and several deaths. The configuration of the surrounding hills shelters the basin from most wind directions, but the valley of the Clyde leaves a gap to the west through which Atlantic gales funnel with considerable force (and unfortunately most gales in the west of Scotland come from the Atlantic).

Although buildings tend to reduce the wind speed, the layout of the streets can be of some importance. In Glasgow the pattern in the older parts of the city tends to be one of east-west arteries running parallel to the Clyde (fig. 97). They allow strong westerly winds to penetrate much further into the city than would be expected.

The average wind speed, as with average temperature and rainfall, has to be used with caution, since it does not tell the frequency or strength of the gusts which often cause most damage. The gusts are partly built into the structure of the wind itself, which is constantly varying in speed. The layout of buildings can cause important local variations, and these can only be examined by making detailed measurements. Gusts of over 65 k.p.h. (40 m.p.h.) can make some outdoor work, especially constructional work, hazardous and, if the average wind is about 50 k.p.h. (30 m.p.h.), gusts of this force can be expected.

Although the number of days when the average wind speed reaches 65 k.p.h. (40 m.p.h.) is small, gusts of this force are quite common, especially between October and March.

The frequency of wind from various directions is also an important climatic factor. It can affect the number of gales, as well as the significance of shelter belts of trees around fields. Fig. 24 shows a wind rose for Glasgow. When this is considered, the importance of the gap between the Renfrewshire and Kilpatrick hills, as well as the one leading to the Forth valley, can be more clearly seen. In both cases the gaps correspond to common wind directions.

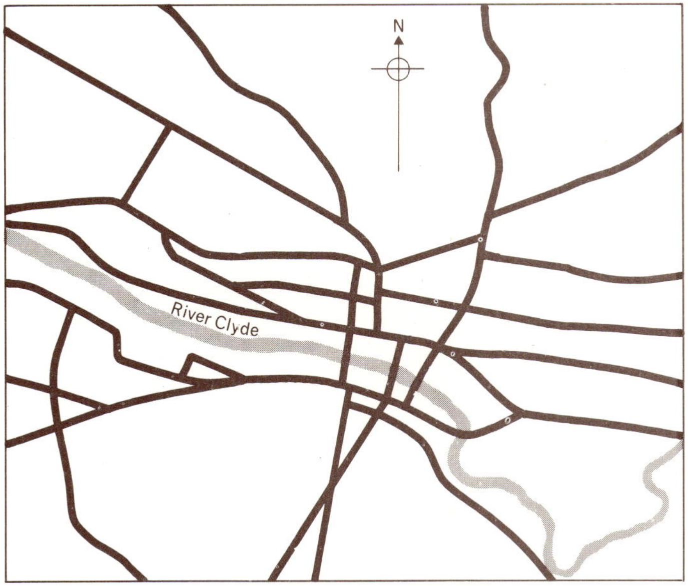

Fig. 97 Street pattern in central Glasgow

School weather station

The study of local climate, especially when its features are related to such things as rural land use, and the layout of a town, makes a very interesting and instructive topic for field work. Unfortunately in many places the necessary local climatic statistics are not available. Much can be learnt, however, from observations made in a school weather station. Although long-term averages cannot be obtained, measurements of day-to-day (and even hour-to-hour) changes in the weather can provide very useful data for a study of the local climate and its significance.

Weather recording of the standard required by the Meteorological Office necessitates the use of very accurate instruments operated under strict scientific conditions which are detailed in the Meteorological Office *Observer's Handbook* (H.M.S.O.). It is possible to make slightly less accurate, though still very useful, observations with a simple and relatively cheap set of instruments. The essential requirements for a simple weather station are:

1 Stevenson screen (a 'do it yourself' construction kit is available)
2 maximum and minimum thermometers
3 grass minimum thermometer
4 wet-and-dry-bulb thermometer
5 rain gauge
6 wind vane
7 barometer

More complicated instruments such as an anemometer, sunshine recorder, or barograph, add considerably to the expense.

With this simple set of instruments it is possible to obtain a fairly accurate picture of local weather. With two sets of instruments at contrasting sites, local variations can be observed.

11 Regional Division of the World

Climate Regions

Although there are many local and regional climatic variations within Britain, the whole country has basically the same climate, and the problems and limitations which it imposes are much the same everywhere. No part of the country has extreme heat in summer or extreme cold in winter. As a result, air conditioning in houses and trains is almost unknown. Similar crops are grown in all parts of the country, and nowhere are hot-climate crops like rice and oranges to be found. On crossing the North Sea or English Channel to northern France, Denmark, or Norway, similar conditions are found. Thus, despite the local variations, the whole of north-west Europe clearly belongs to one climatic region.

On travelling further afield, for example, to Greece or Russia, very different conditions are experienced. Moscow has much more extreme temperatures than any part of western Europe, while the Athens area is mild enough for the cultivation of crops throughout the year. Each of these places clearly belongs to a different climatic region. When the climatic variations throughout the world are examined, however, it is soon noticed that the number of varieties is quite small, and that similar climates are found at places thousands of miles apart. For example, coastal British Columbia and New Zealand have weather very similar to that found in north-west Europe; hot deserts similar to the Sahara occur in northern Chile and western Australia. It is therefore possible to divide the world up into a number of climatic regions and, although various schemes for making this division have been worked out, they are all very similar. The main cause of the differences is that there is no well-defined line separating one climatic region from another—rather a zone of change—and this allows considerable latitude when deciding where to draw a line on a map.

The variety of climate types is of great value to man, as it gives rise to the many species of plants, both natural and cultivated. Moreover, the various environments are available for development in many different parts of the world. This latter feature was of great importance during the Age of Discovery, when Europeans found climates similar to those to which they were accustomed, in places as widely scattered as Canada, Brazil, and Australia.

The division of the world into the climatic regions shown on fig. 98 is based on the following principles:

1 TEMPERATURE

The general decrease in temperature from the equator to the poles has been used to divide the world

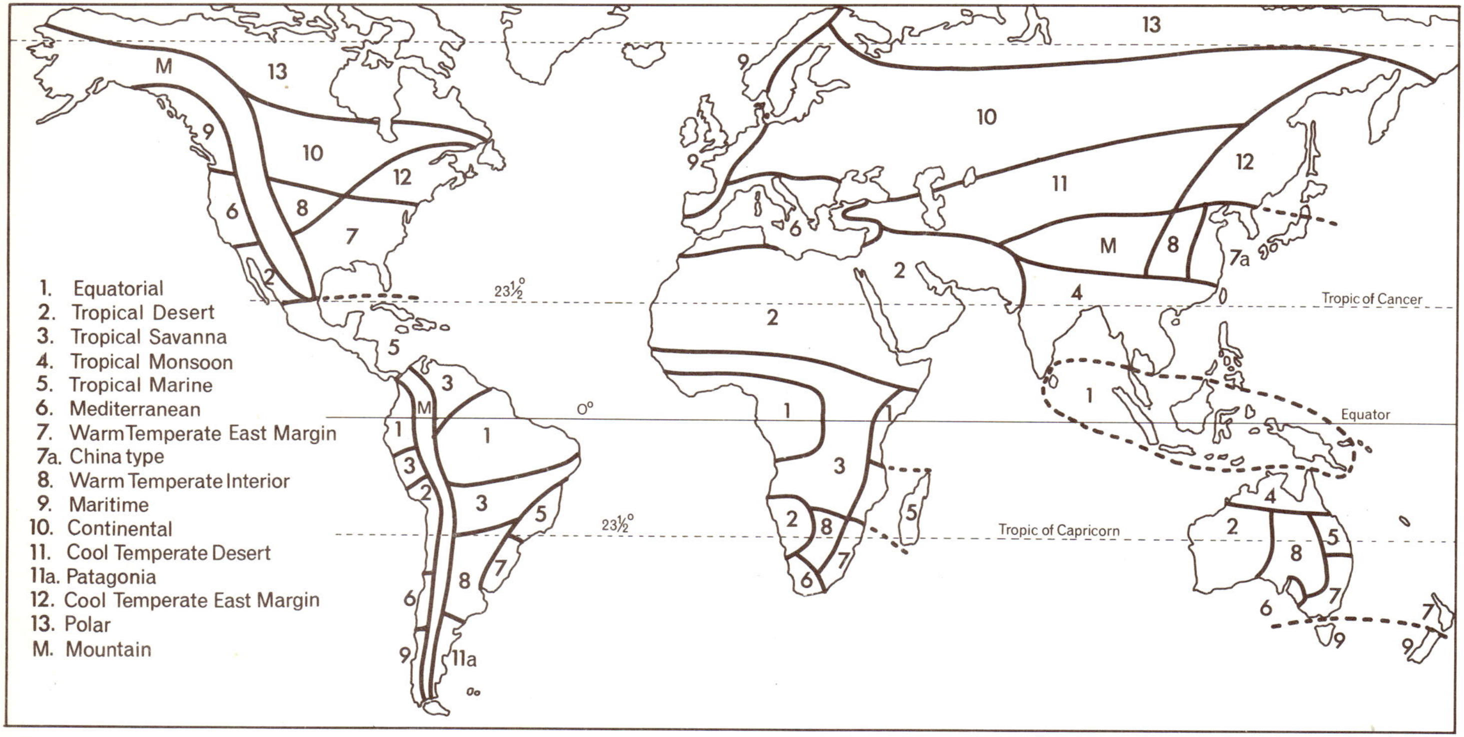

Fig. 98 Natural regions of the world (a) Climatic regions
(numbers on climate map correspond with paragraph numbers in Section C)

up into the following three temperature zones (fig. 99 and 72).

(*a*) **Tropical** (extending beyond the tropic lines) where there is no cold season.

(*b*) **Temperate** where there is a clearly defined hot (summer) and cold (winter) season. This region contains a sufficient variety of temperatures to justify subdividing it into a **warm temperate** and a **cool temperate zone**.

(*c*) **Polar** where it is cold for most of the year and, although temperatures may rise fairly high in summer, they only do so for a very short period.

2 RAINFALL

The basic temperature zones are subdivided according to variations of rainfall regime into:

(*a*) Regions with no dry season.

(*b*) Regions with a dry season and a wet season.

(*c*) Regions of drought.

3 ALTITUDE

The climates of mountains and high plateaus often differ significantly from those of neighbouring lowlands, and accordingly are treated separately.

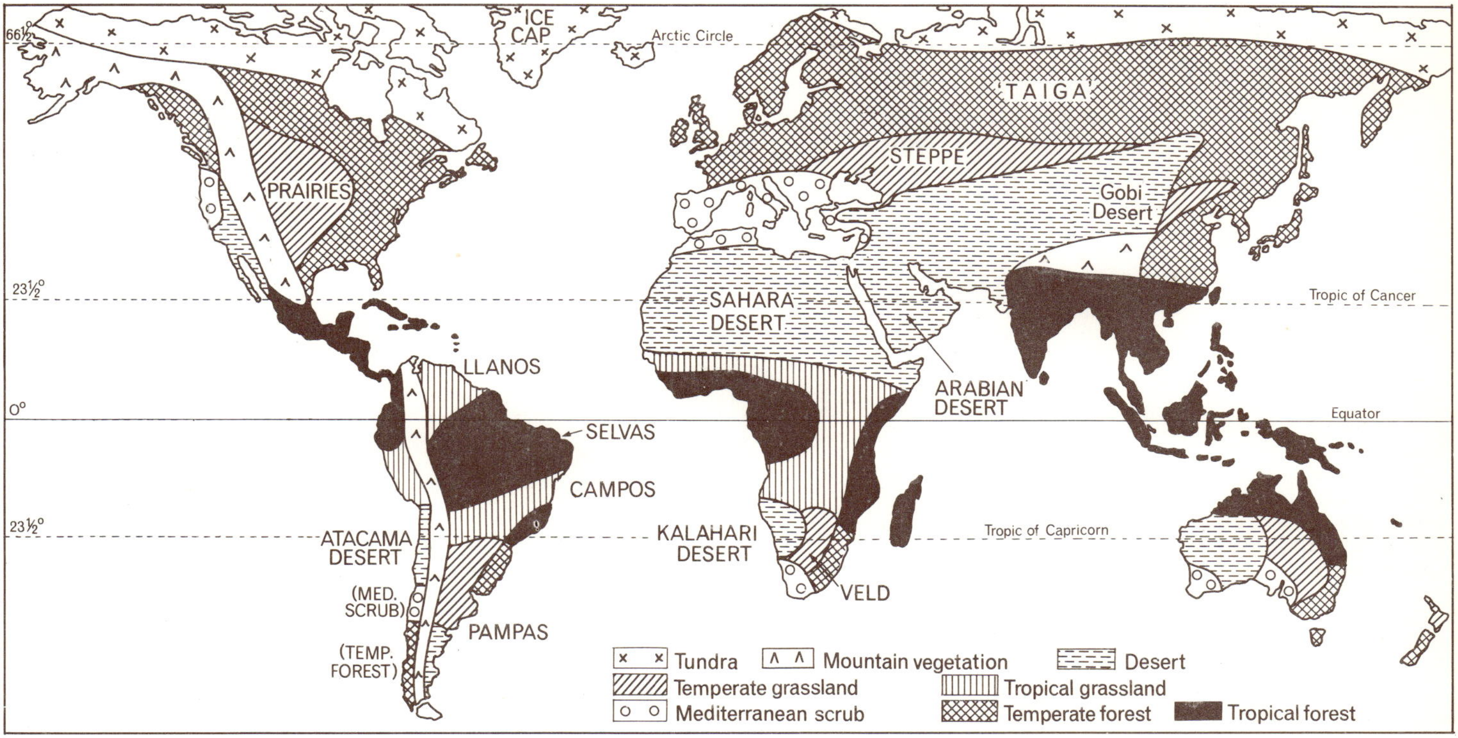

Fig. 98b Natural vegetation

In the following chapters each climate region (sometimes called a *natural region*) will be considered in turn, along with a brief consideration of its vegetation (natural and cultivated) and the way of life of the people.

Natural Vegetation

All plants have limiting temperature requirements and, if the temperature rises above or falls below these limits for any length of time, the plant will cease to grow. In areas of large temperature differences between summer and winter, plants adapt themselves to these conditions in a variety of ways.

1 They may have a short life cycle, ending in the production of seeds which can survive the cold winter, and produce next year's plants in the spring.
2 The leaves may be shed or wither during the winter, while the plant is dormant, and new leaves are produced next spring when growth begins anew.
3 They may have hard leaves which survive the winter cold and are ready to assist plant growth as

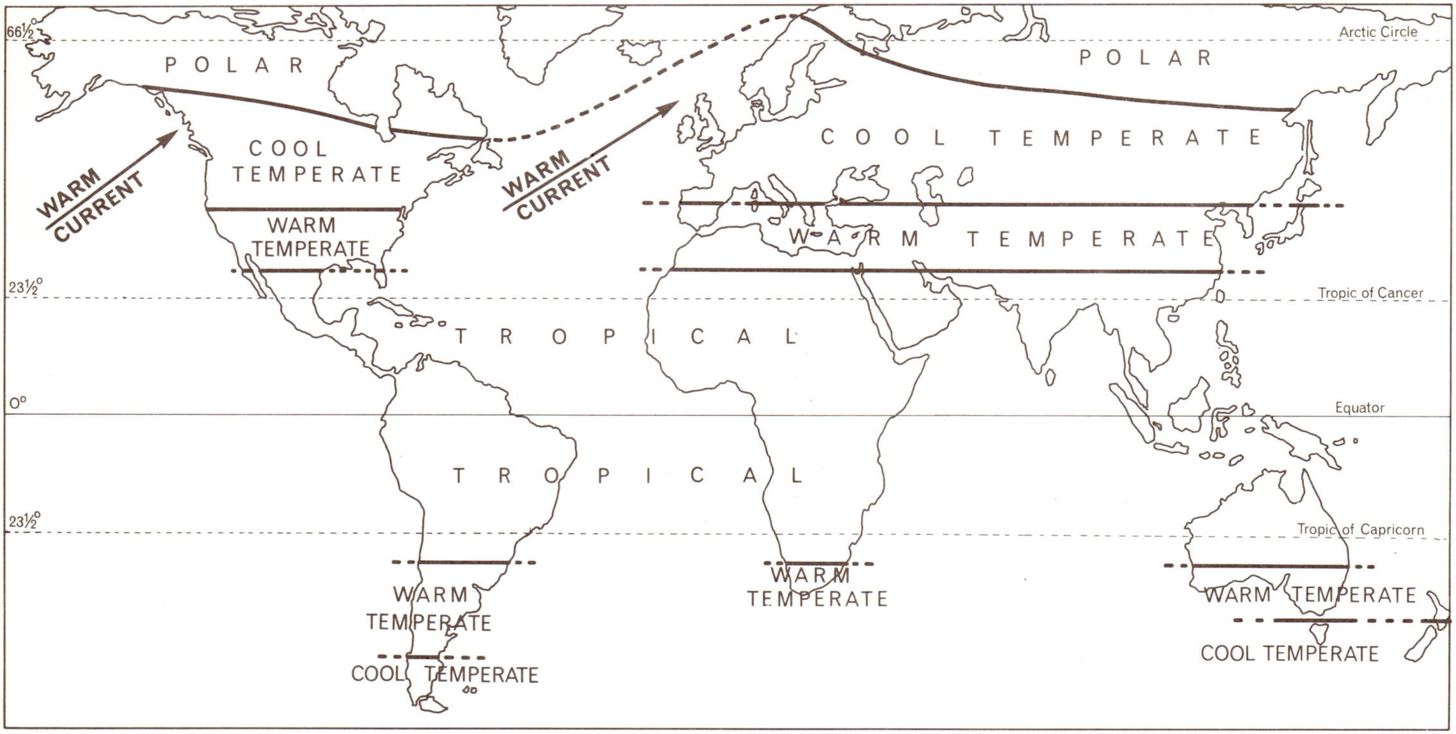

Fig. 99 Temperature regions of the world

soon as warmer weather comes. These are often found in areas where the growing season is rather short.

All plants require a certain amount of water and, as a general rule, the wetter areas have a greater vegetation cover than drier areas.

FORESTS

Trees require more water than most other plants. Much of the moisture taken in through the roots is lost by transpiration from the leaves. In tropical latitudes, at least 750 mm (30 inches) annual rainfall is required for the development of forests; but in cooler temperate areas 500 mm (20 inches) is sufficient if winters are cold enough to reduce the rate of transpiration sufficiently, and if there is a summer maximum of rainfall. In areas where tree growth is interrupted by a dry season (Mediterranean areas) or cold winters (maritime areas), the trees tend to lose their leaves at the same time. At the equator, where neither drought nor cold hinder plant growth, there is no single season of leaf fall.

GRASSLANDS

Although grass is often present in the undergrowth of forested areas, it becomes dominant only where the rainfall is insufficient for trees. The term *grass* covers a number of non-woody plants called *herbs*, some of which propagate annually from seeds, while others do so by sending up new shoots from old roots. All of the grasses have shallow roots and, although they require a regular supply of water during the growing season, they can survive periods of adverse conditions, such as severe cold, drought and fire.

In many parts of the world the natural vegetation described in the following chapters has been modified or removed by man. For example, the moorlands of the Scottish Highlands have resulted from the removal of the old forest cover and decades of sheep grazing; in many of the temperate grasslands—particularly in the southern hemisphere—the native grasses have been replaced by new varieties grown from imported seeds.

Way of Life

The way of life of the people in the various natural regions is a result of many complex factors, of which the climate and vegetation are only two. Their influence on the people is most clearly seen in rural areas, where there is an obvious connection between the climate and the type of farming practised. In general, farming has developed on the natural grasslands where annual rearing and ploughing are relatively easy, while the forests, with the problem of clearing the trees, have been little used for farming unless—as in western Europe—pressure of population has made the clearance essential.

The section in the following chapters dealing with man's activities in the various natural regions, gives some indication of the influence climate has had on them. In order to understand the complexity of man's response to his environment reference must be made to regional textbooks and articles in geographical journals dealing with specific areas.

STUDY SECTION FOR CHAPTER 11

1 What is a climatic region?

2 For each of the three major temperature regions of the world:
(*a*) give its name,
(*b*) describe briefly its main characteristics.

3 What are the three major rainfall regions found in the world?

4 Trace Fig. 98a and shade the regions as follows:

if they have no dry season.
if they have a dry season in summer.
if they have a dry season in winter
if they are areas of drought.

(Refer to General Features sections at start of Chapters 12, 13, and 14.)

Tropical Climates

12

General Features

Tropical climates occur between the mid-latitude high-pressure belts which keep out the cold air and disturbed weather of temperate latitudes. Since the air is uniformly warm, temperature variations are mainly caused by variations in the angle of incidence of the sun's rays (caused by the movement of the overhead sun). When two air masses meet, as at the **intertropical front** (see fig. 26), their physical characteristics of temperature and humidity are sufficiently similar for the depressions which form to be much less active than those occurring along the **polar front**. Since the high-pressure belts are situated on the poleward side of the tropic lines, the climates of this type extend a varying distance beyond $23\frac{1}{2}°$ latitude, and their limits are partly governed by the local configuration of the land.

The five tropical climates all have high average annual temperatures. Although the temperature tends to decrease as distance from the equator increases, there is no cold season, so that cultivation can take place throughout the year. Rainfall does vary greatly, however, both in annual total and seasonal distribution to give the following climate divisions:

1 Equatorial climate with heavy rainfall and no dry season.
2 Tropical desert climate with very light and unreliable rainfall.
3 Tropical savanna climate } with summer rain
4 Tropical monsoon climate } and winter drought.
5 Tropical marine climate, with no dry season but greater variation in the annual rainfall than in equatorial areas, and rather more rain in summer than in winter.

Although the terms *summer* and *winter* are used, it should be remembered that these seasons do not exist in the tropics in the same sense as they do in temperate latitudes. They are used here simply as convenient terms for describing a particular time of the year. Throughout much of the tropical world it is the distribution of rainfall which is of most significance for dividing the year into its seasons.

1 Equatorial Climate

LOCATION

Equatorial climate is found in a belt extending about 5°–10° latitude on either side of the equator (fig. 100) and it is most fully developed in:

The Amazon Basin
The Congo Basin and coastal West Africa
The East Indies

FACTORS

In these areas the sun is never far from overhead at

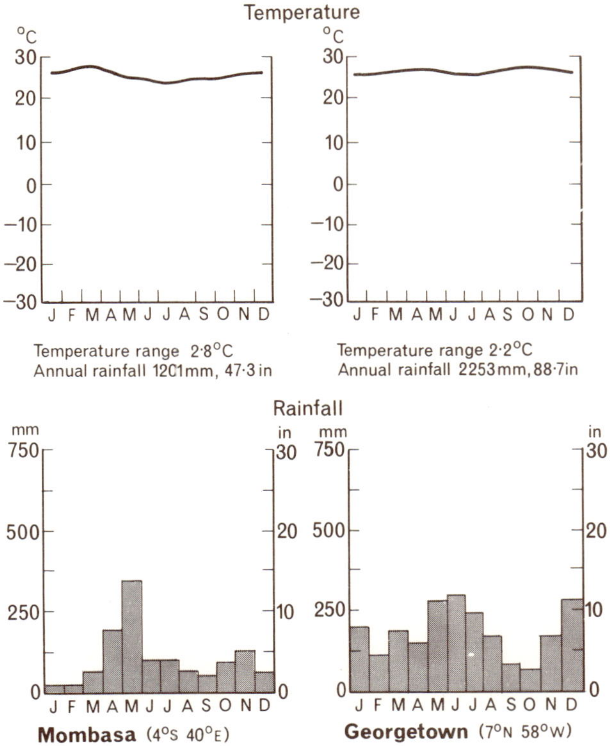

Fig. 100 Two equatorial stations

noon. As a result it is hot, with very little variation in temperature throughout the year. The length of day also varies only slightly, with sunrise about 6 am and sunset about 6 pm, and very little twilight.

The heat causes strong convection currents. Combined with the cyclonic disturbances and rising air along the intertropical front, these currents result in heavy and well-distributed rainfall.

In this region of low pressure and rising air between the trade-wind belts, winds are seldom strong; it is sometimes known as the **doldrums**.

TEMPERATURE

The average monthly temperatures seldom vary more than about 5°C from 25°C. The diurnal range is also small, although it may be as much as 10°C.

RAINFALL

There is a greater local variation in rainfall than in temperature. Annual totals are usually between 1500 mm (60 inches) and 2000 mm (80 inches). Although there is no dry season, there are considerable differences in the monthly fall at any one station. Very often the heaviest rains are immediately after the passage of the overhead sun, and much of the rain is of the convectional type, falling in the afternoon when the ground is hottest.

NATURAL VEGETATION

Tropical rain forests occur in hot wet regions with no dry or cool season to hinder plant growth. The trees of these forests are broad-leaved evergreen varieties which grow to heights of over 30 metres (100 feet), and the intermingling of their leaves and branches produces a canopy which shuts out much of the light from the lower layers. Many climbing rope-like plants called *lianas* twine round the trunks and branches of the trees. Below the canopy, some smaller trees and shrubs grow, but this undergrowth is restricted by the lack of light. When a break occurs in the canopy, such as at an old clearing or along the banks of a river, the undergrowth increases greatly, giving a form of jungle which is very difficult to penetrate. In these forests the local climate is very humid, calm and dull. Europeans find it most unpleasant and unhealthy. A large number of different species of trees grow in these forests, but only a few, such as mahogany and rosewood, are of commercial value.

WAYS OF LIFE

The equatorial forests are generally sparsely populated, with the exception of the island of Java. Most people make their living by some form of cultivation. The absence of large areas of grassland, and the prevalence of disease in the hot damp climate, make animal rearing almost unknown. The only animals found in any numbers are work beasts like the water buffalo which is used for drawing ploughs in rice fields. Large wild animals are also notable by their absence, and the forests are inhabited by small creatures like snakes and lizards as well as numerous birds.

The main crops of the region are manioc (a root crop similar to parsnip in appearance, which is ground into a powder to make flour), sweet potatoes, and rice. Before cultivation can take place, a clearing has to be made in the forest; the ground between the tree stumps is loosened by a digging stick or hoe. The crops grow well for a few years on the fertility of the decayed leaves, but the heavy rainfall causes rapid leaching. Since animal manure is not available, the soil soon loses its richness. Chemical fertilizers in powder form would soon be washed away and, in any case, the farmers do not have the money to buy them. When the yield from the clearing deteriorates, it is abandoned and a new clearing is made. Much of this farming, known as **shifting cultivation**, is done by village communities rather than individual farmers; when the fertility of one area has been exhausted, the whole village may move to a new site. The forest gradually reestablishes itself on the old clearing, and in time (possibly 20 years) the leaf mould will refertilize the soil. This type of farming requires a large area of land to support a relatively small number of people and is one of the reasons for the sparse population of equatorial areas.

There are many trees bearing useful fruits and nuts which are collected to supplement the diet. In West Africa the value of the fruits has been turned to commercial advantage by the villagers who, in addition to their subsistence farming, produce cocoa and palm oil on plantations near the villages. These products, unlike the subsistence crops, are in demand on the world market, and their sale adds considerably to the wealth of the area. In Malaysia large tracts of forest have been cleared for rubber plantations which are owned by commercial companies who employ many of the local inhabitants. Although the plantation workers have small patches of land where they grow rice for themselves and their families, they are basically employees dependent on the prosperity of the plantation rather than subsistence farmers.

Although the equatorial regions contain some of the most extensive forests in the world, they are not very important for the production of commercial timber. The main reasons for this are:

(*a*) The trees do not grow in pure stands, and the commercially valuable ones are scattered haphazardly through the forest.

(*b*) The great girth of many of the trees, the hardness of the wood, and the hot humid climate make felling extremely hard work.

(*c*) The thick undergrowth near rivers and clearings and soft ground make the movement of logs difficult.

(*d*) The demand for hardwood on the world market is very limited.

2 Tropical or Hot Desert Climate

LOCATION

Tropical or hot desert climate is found at the northern and southern limits of the tropical region on the west side of large land masses (fig. 101). It normally extends on both sides of the tropic lines. The main deserts are:

Atacama in South America.
Kalahari in South Africa.
West Australian in Australia.
Mexican and Californian in North America.
Sahara in Africa which continues into Asia as the Arabian, Persian, and Thar.

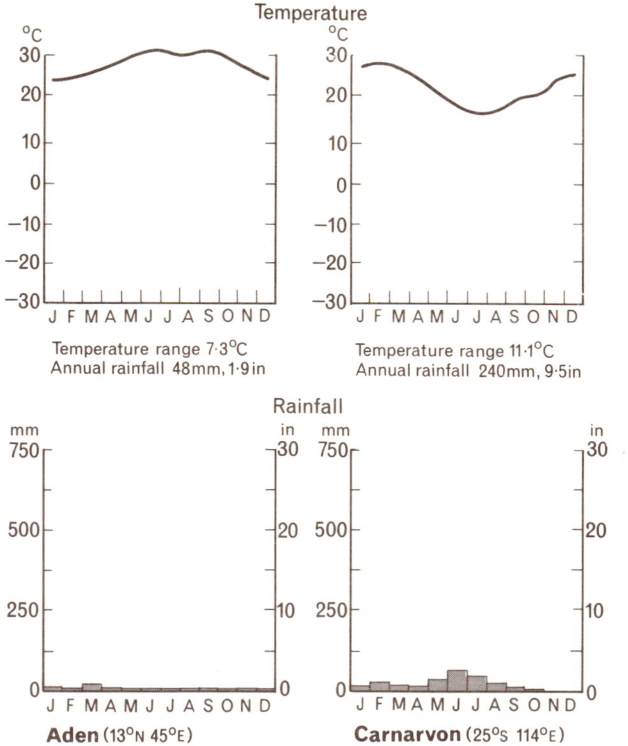

Fig. 101 Two tropical desert stations

FACTORS

The excessive drought of these areas results from two main factors.

1 They are located where the mid-latitude high-pressure belt gives very stable atmospheric conditions.

2 The air movement is from the east, and this air is dry as a result of passing over large areas of land (as in the case of the Sahara) or over very high mountains (as at the Atacama).

The clear skies and high altitude of the sun result in very high temperatures, and some of the hottest parts of the earth are located in the deserts.

TEMPERATURE

Average monthly temperatures range from about 15°C to about 30°C. This range, which is larger than for equatorial regions, results partly from the greater variation in the angle of the sun's rays in the course of a year, and also from the much greater diurnal range. The difference between day and night temperatures may be as much as 20°C, since the bare ground heats rapidly during the day and loses its heat just as rapidly into the clear atmosphere at night.

RAINFALL

In hot parts of the world an average annual rainfall of under 250 mm (10 inches) usually gives desert conditions, but this average figure can be very misleading as the amount varies greatly from year to year. Very few parts of desert areas never have any rain, but it occurs in torrential local downpours which may be separated by months and even years. General rain is unknown.

NATURAL VEGETATION

Semi-arid vegetation occurs in regions where drought is a problem, and the plants are adapted to withstand drought in a variety of ways.

1 There are many annual plants which mature rapidly from seed when there is some moisture, and then they produce seed which lies dormant till the next rain.

2 Bulbs and tubers remain dormant in the soil when it is dry, but grow rapidly as soon as it rains.

3 Trees and shrubs may have long tap roots to reach the water table. Cells for storing water are present in thick leaves with hard glossy surfaces which may have spines. These plants are often rather small, with twisted woody trunks and

branches, and are well spaced out so as not to overtax the available water supply.

Scrub and thorn forest occurs on the drier fringes of some savanna areas and is particularly well developed in the Gran Chaco of South America.

WAYS OF LIFE

The shortage of water for plants, animals and people makes the desert environment a very difficult one, thus these areas are among the most sparsely populated in the world (with the exception of polar and very high mountain regions). The traditional way of life in many deserts is nomadic pastoralism. Although some areas are completely devoid of vegetation, there are places, especially on the fringes, where a poor form of coarse pasture is found. This is used as grazing for sheep, goats, and camels. Great care has to be taken to avoid overgrazing this pasture, and the tribes with their animals are constantly on the move from one grazing area to another. This movement is not haphazard. Each tribe has its own territory, but a very large area of land is required to support a small number of people. The animals provide food for their owners, and at market towns like Kano on the edge of the desert hides and wool are exchanged for other goods. The Bedouin of the Sahara is one of the best known nomadic tribes of this type.

When water is available from rivers or springs, however, desert areas are extremely fertile, and particularly so when river water contains silt to enrich the soil. Some of the earliest civilizations developed in the desert lands of Mesopotamia and Egypt, where the great rivers Tigris and Nile, fed by rain falling well beyond the desert regions, provided irrigation water. Although an irrigation system requires skilled engineering works and a stable sedentary civilization to organize the supply of water, it makes cultivation easier than in areas dependent on rainfall, as the water can be supplied to the fields in the quantity and at the times most favourable to the crops. The high temperatures and abundant sunshine of the deserts allow two or three harvests a year from the same piece of land; as a result, irrigated areas are often very densely populated. Many food crops such as rice, wheat, sugar cane, fruit and vegetables are grown; and in some areas, such as Egypt, cotton is grown as a commercial crop.

In the desert oasis, underground aquifers supply the water and the date palm is the most common plant. As well as providing highly nutritious fruit, these trees give shade for the cultivation of other crops.

The sunshine and drought have also had an influence on local architecture. Many desert settlements consist of mud-brick houses with small windows and flat roofs.

3 Tropical Savanna Climate

LOCATION

Tropical savanna climate is found immediately north and south of the equatorial areas of Africa and South America (fig. 102). It does not occur in Asia or Australia, where it is replaced by the monsoon climate, which, although similar, has a number of distinctive features. The East African plateau (Tanzania, Kenya, Uganda) also has a form of savanna climate, even though it is mostly in equatorial latitudes.

FACTORS

On one side of the savanna region lies the equatorial low pressure with its unstable air and doldrum rain, while on the other there is the mid-latitude high-pressure area of stability and drought. These two pressure belts with their characteristic weather move north and south with the annual migration of the overhead sun (fig. 103). As a result the savanna areas

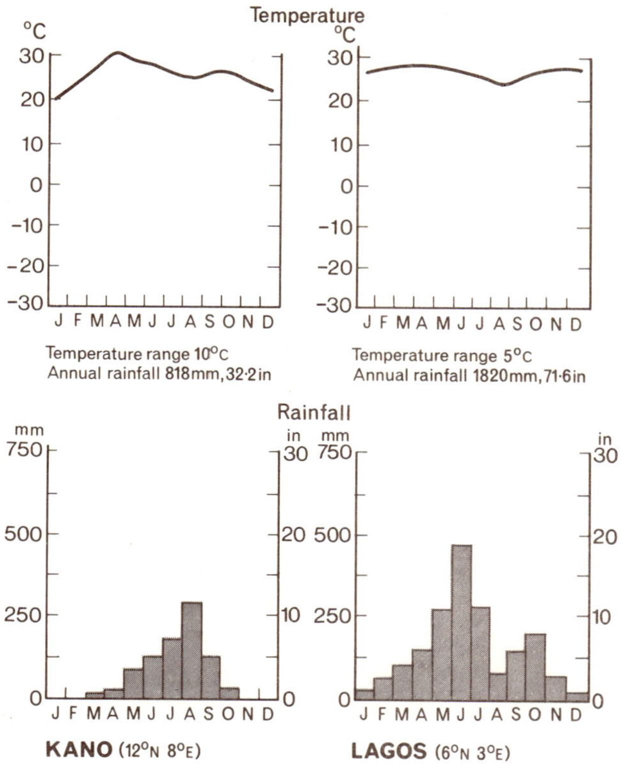

Fig. 102 Two savanna stations

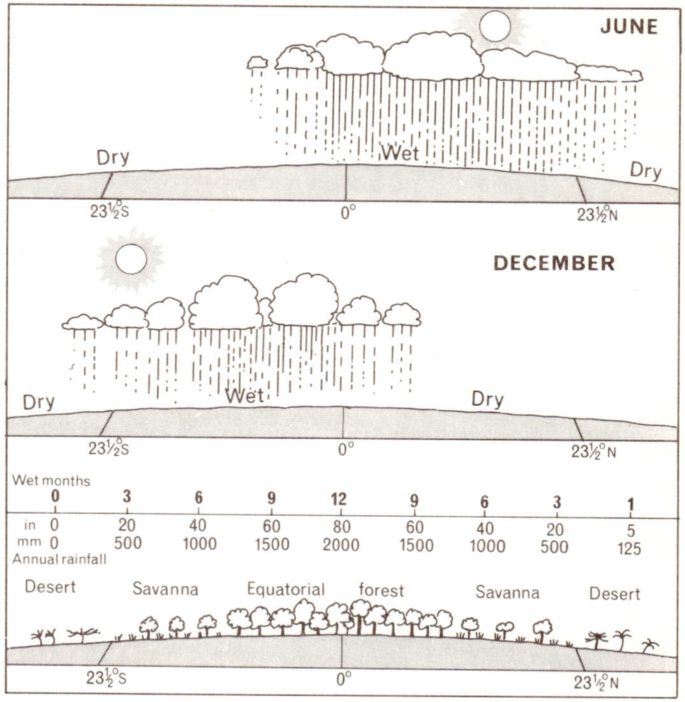

Fig. 103 Migration of doldrum rains with overhead sun affects climate and vegetation

have wet conditions in 'summer' and dry weather in 'winter'.

Although the dry season is in 'winter', temperatures are not low, since skies are clear and the midday sun is not far from the vertical. The low humidity and fresh breezes of the dry season make it much more pleasant and healthy than the wet season with its cloudy calm humid conditions.

TEMPERATURE

Temperatures are intermediate between those of equatorial and desert areas and are at their highest just before the start of the wet season. The cloud cover causes a slight drop in temperature in midsummer. Although the diurnal range is greatest in winter, the night temperatures never drop as low as they do in desert areas. Because of its equatorial latitude, the annual range on the East African plateau is less than in most other savanna areas.

RAINFALL

Most rainfall is of the convectional or frontal types associated with the equatorial region. The length of the wet season decreases with increasing distance from the equator, and there is a corresponding decrease in the annual rainfall, so that totals can vary from about 1500 mm (60 inches) to as little as 375 mm (15 inches).

NATURAL VEGETATION

Savanna grasslands occur in tropical regions with marked wet and dry seasons. They lie between the tropical rain forests and the deserts, so that the term *savanna* covers a gradual transition from thin forest with grass undergrowth, through extensive grasslands with scattered trees (sometimes called parkland), to thin short grass as the annual rainfall decreases and the dry season lengthens.

The grasses of these areas grow in tufts separated by bare earth, and they are coarse in texture. In the wetter areas the grass is very tall, sometimes reaching heights of over 3 metres (10 feet). As rainfall decreases, the grass is shorter and the tufts are more widely spaced. The grass normally reaches its greatest height towards the end of the wet season, when seeds are produced. As the dry season progresses, the grass turns brown and dies down till the start of the next wet season, when the new seeds germinate and growth starts afresh. The savanna grasslands are not well suited for use as grazing areas for good-quality cattle, and the seasonal withering during the dry season is an additional problem in developing these areas for pastoral farming.

The trees of the savanna areas are deciduous, losing their leaves in the dry season, and they are of little commercial value for their timber.

WAYS OF LIFE

The tropical savanna is noted for an abundance of wild life. This is particularly true of Africa, where antelope, giraffe, zebra, lion, and elephant are numerous. In the Australian and South American savannas there are fewer species of native animals, but of course this is true of all parts of these continents.

Cattle rearing from villages or isolated cattle stations is the most common way of life on the savanna grasslands, but they are not important suppliers of beef or dairy products to the world markets. The reasons for this are varied. In Australia and South America very few of the indigenous native peoples lived on the grasslands. European settlers were attracted to other parts of the continents, so that the savanna lands are still largely awaiting development. Africa, on the other hand, has a large number of native cattle-herding tribes like the Masai of East Africa, but these people breed a humped zebu type of animal which gives a poor milk yield and little good-quality meat. They provide food for their owners, but are quite unsuitable for commercial cattle farming. There are a number of environmental difficulties which make the introduction of pedigree dairy and beef animals very difficult. Good-quality animals like the Ayrshires and Herefords which have been bred in temperate latitudes do not easily adapt to the high tropical temperatures, and lack of water is a great problem. Not only is there a long dry season but the rains of the wet season are very unreliable, often varying greatly from the average. The savanna grass is quite unlike the lush green pastures of Britain, being tall and coarse during the wet season, and then drying up and withering when the rains cease. Disease is another problem; in particular sleeping sickness, which is spread by the tsetse fly. This insect is found wherever there are wild animals and, although energetic measures are being taken to eradicate it, there are still only a limited number of tsetse-free areas.

Cultivation is largely restricted to the African savannas where once again the weather presents problems. Unless irrigation water is available, crops can only be grown during the wet season, and the unreliable rainfall often causes failure of the harvest.

When the rains do come, they are frequently very intense, and rapid soil erosion of the newly dug, or ploughed, fields is common. A variety of crops including maize, manioc, and millet are cultivated for local use, but there are few commercial crops for the world market. Groundnuts from West Africa, and coffee and sisal from East Africa, are probably the most important.

As already mentioned, the savanna areas of Australia and South America are sparsely populated. Although many more people live on the African grasslands, they are not densely inhabited. Undoubtedly all these areas could support larger populations, but there are many environmental problems which are at present not fully understood, far less overcome.

4 Tropical Monsoon Climate

LOCATION

Tropical monsoon climate is most fully developed over the 'Indian subcontinent' and parts of south-east Asia, including Burma, Thailand, and 'Indochina', but it also occurs with less intensity in northern Australia (fig. 104). Some parts of the Guinea coast of West Africa also have slight monsoon characteristics.

FACTORS

In winter, the monsoon areas are affected by high pressure and trade winds, as are the savanna areas; but in summer, low pressure over the land reinforces the low pressure brought by the doldrum belt, and very strong wet onshore winds bring torrential rain and thunderstorms. These conditions develop when a large land mass lies on the poleward side of a large sea area. Although it is sometimes difficult to distinguish between monsoon and savanna from temperature and rainfall graphs, the monsoon has three

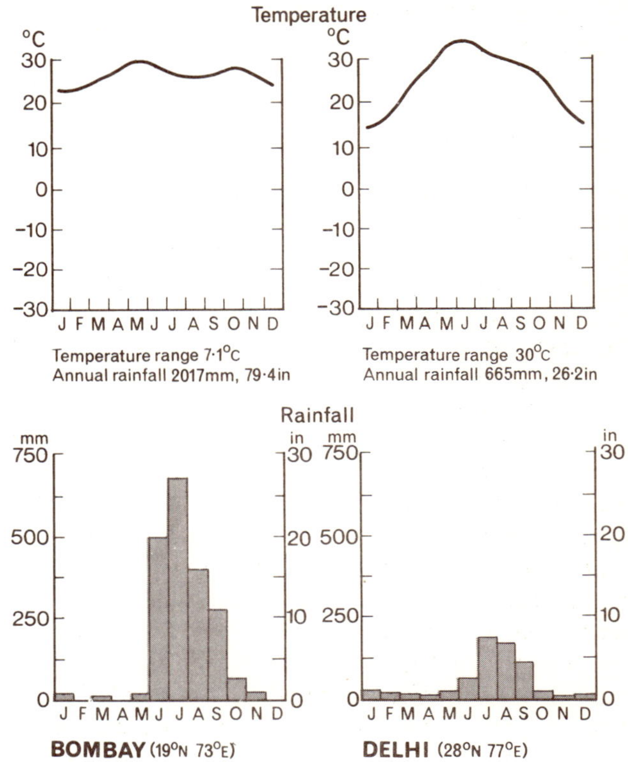

Fig. 104 **Two monsoon stations**
Note large annual temperature range and lighter rainfall at Delhi resulting from inland situation and greater latitude.

characteristic features which are not found in savanna areas.

1 Reversal of wind direction accompanying the change from dry to wet season.
2 Stormy conditions during the wet season.
3 Sudden onset of the rains which is sometimes described as the 'bursting' of the monsoon.

TEMPERATURE

The year is normally divided into:

(*a*) Hot dry season—spring and early summer.

(*b*) Hot wet or monsoon season—summer and early autumn.
(*c*) Cool dry season—winter.

The temperatures are very similar to those of savanna areas.

RAINFALL

Although the annual rainfall pattern is similar to that of savanna areas, distribution over the land is affected to a much greater extent by relief features, since there is a strong wind bringing the moist air from the ocean. Windward slopes of hills are very much wetter than leeward slopes, and there is a noticeable decrease in rainfall inland from windward coasts. This pattern is well illustrated in India (fig. 105), with the west coast of the Deccan much wetter than the east, and rainfall decreasing up the Ganges valley.

Another characteristic of monsoon rain is its intensity. A wet day in India may produce over ten times as much rain as would fall in a wet day in Britain. One of the wettest places in the world, Cherrapunji, with an average annual total of 11,600 mm (457 inches), is in the Indian monsoon region.

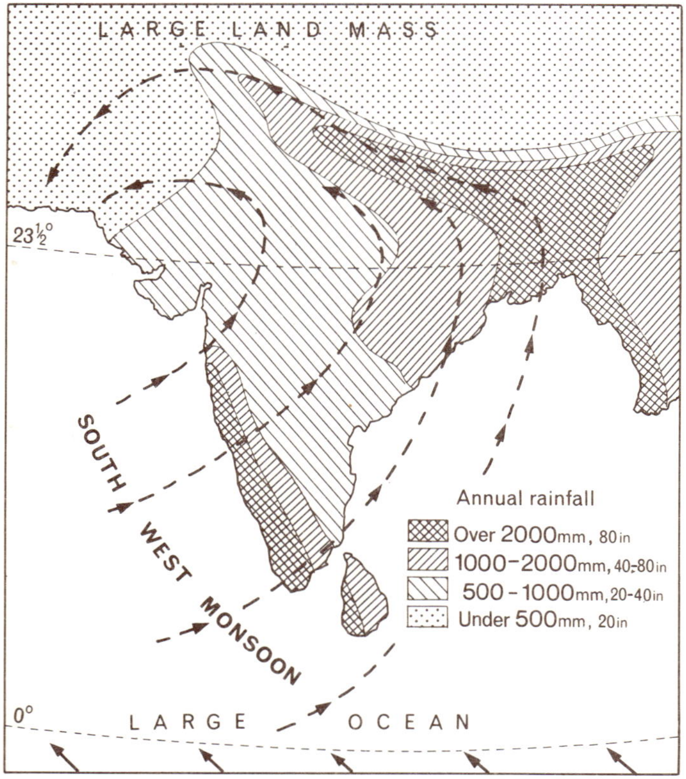

Fig. 105 (*a*) **Winds over 'India' in summer,** (*b*) **annual rainfall**

NATURAL VEGETATION

On the fringes of the equatorial regions and in the wetter monsoon areas, the evergreen forest gives way to one with a lower and more open canopy. This results from the presence of a short dry season, slowing growth for part of the year, and the presence of some deciduous trees which lose their leaves during the dry months. More light can penetrate this forest, and as a result the undergrowth is often very thick, although it tends to die down when the rains cease. Acacia and teak are common trees.

WAYS OF LIFE

The monsoon regions contain some of the greatest population density to be found anywhere in the world. These very dense areas are on alluvial low-lands along the courses of rivers like the Ganges and Mekong, and the people are mostly farmers engaged in subsistence agriculture. The flat land, rich alluvial soil, heavy monsoon rains, and high temperatures make the fields very productive. The staple food is rice which grows in embanked paddy fields. At the start of the wet season rice is planted in the flooded fields, and it is ready to be harvested about four months later when the rains stop. Cultivation continues during the dry season, when irrigation water is obtained from rivers, wells, and tanks (small reservoirs in which some of the monsoon rain is stored).

There is not much animal rearing, but the water buffalo assists with ploughing and the turning of the water wheels for irrigation. In addition to rice, a great variety of other crops, such as oil seeds, sugar-cane, maize, and millet, is grown and on the hill slopes, which are often terraced, tea is a common crop. Most of the farm produce is used locally, but tea, and jute from the Ganges delta, are exported.

Not all parts of the monsoon lands are densely populated, however. The hill areas, which are densely forested and where communications are difficult, contain very few people; and the drier lowlands like the Punjab are only densely populated where major irrigation schemes associated with rivers like the Indus have been developed.

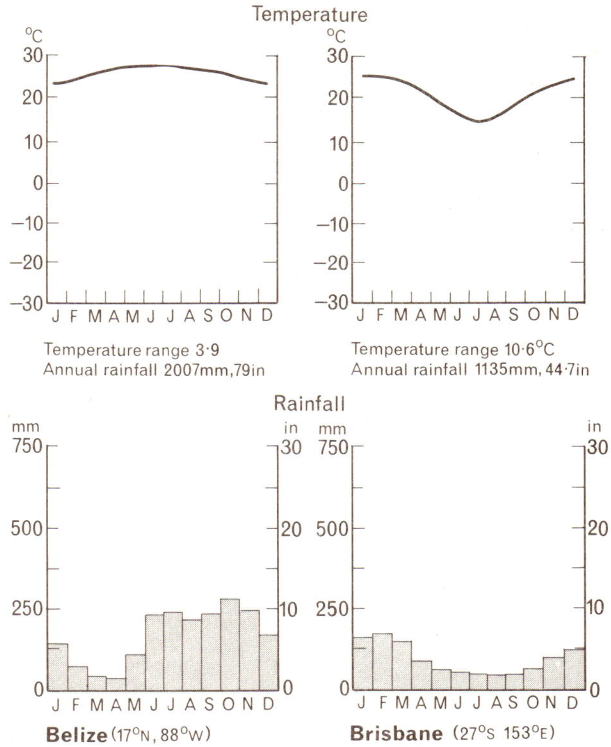

Fig. 106 Two tropical marine stations

5 Tropical Marine or East Coast Climate

LOCATION

Tropical marine climate is found on the east coasts of continents in similar latitudes to the desert areas (fig. 106). The principal regions are:

The West Indies and Central America.
Coastal Brazil south of Recife.
Mozambique and the east side of Madagascar. (Northern Mozambique is in the Madagascar rain shadow and is drier, especially in winter.)
Coastal Queensland south of Cairns.

FACTORS

Tropical marine climate, like that of savanna areas, is affected by the annual migration of the pressure belts. In summer it is under the influence of the equatorial low pressure with its doldrum rains and light variable winds, while in winter the trade winds associated with the mid-latitude high pressure are dominant. Since these winds have blown across considerable areas of sea, they give rain, but this time of the orographical type, and the humid summer conditions are replaced by much fresher weather.

TEMPERATURE

Temperature is fairly high throughout the year, but the range is greater than that of equatorial areas, being very similar to that found in savanna and monsoon climates.

RAINFALL

There is no dry season but, while the summer rains are fairly evenly distributed, those of winter are greatly affected by the relief of the land. As a result the annual totals vary greatly from place to place. This is well illustrated in Jamaica and Central America, as shown on fig. 107. Most stations have rather more rain in summer than in winter.

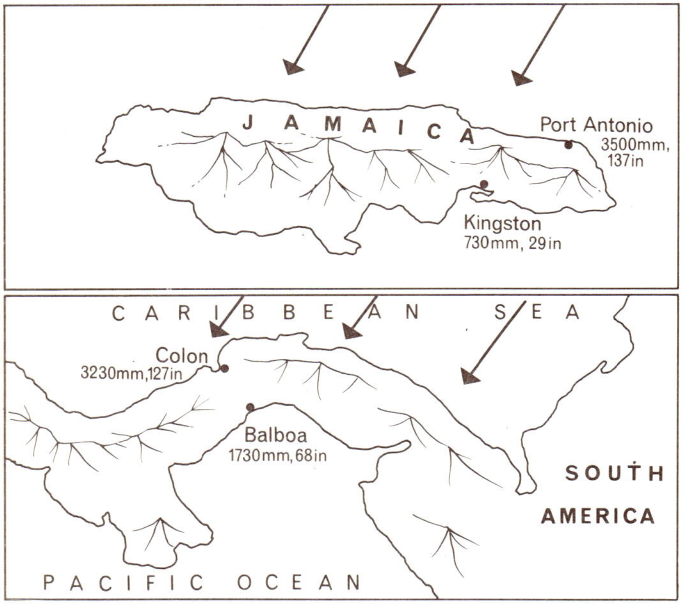

Fig. 107 Rainfall differences between windward and leeward coasts of tropical marine areas

NATURAL VEGETATION

The abundant rainfall and relatively high temperatures result in luxuriant hardwood forests with fairly dense undergrowth, which in many ways resemble the monsoon forest. Many types of tree grow in these forests; some of the better-known species are walnut, evergreen oak, and eucalyptus.

WAYS OF LIFE

Climatic conditions are favourable to agriculture. Much of the forest has been cleared for small farms on which crops like rice, sugarcane, coffee, cotton, bananas, and pineapples are grown. Although most cultivation is for the local market, a number of areas have been developed as plantations catering for world trade. Among the more important plantation areas are the West Indies (sugar and bananas), Brazil (coffee and cotton), and Queensland (sugarcane and fruit). Population density is generally higher than in the savanna areas, partly as a result of a more favourable climate, and partly because the coastal situation has encouraged the growth of ports such as Recife, Brisbane, and Durban which handle trade from a hinterland which extends well beyond the coastal strip. In some of the West Indian islands the attractive climate and extensive beaches have resulted in the growth of tourism as a new factor in the economic life of the area.

STUDY SECTION ON CHAPTER 12

1 Why do all tropical climates have hot or warm weather throughout the year and no cold season?

2 What is used instead of temperature differences for dividing the year into seasons in the tropics?

3 For equatorial regions give:
(*a*) two reasons for the heavy rainfall,
(*b*) the wettest time of the year,
(*c*) the wettest time of the day.

4 State the two main causes of drought in desert areas.

5 Why do desert areas have a larger temperature range (both annual and diurnal) than equatorial areas?

6 What are the two main air masses affecting savanna areas? For each one:
(*a*) state the time of year when it comes.
(*b*) describe the type of weather it brings.

7 Name the main parts of the world with a monsoon climate.

8 State three ways in which monsoon climate differs from savanna climate.

9 What are the main differences between 'summer' and 'winter' in places with a tropical marine climate?

10 Identify each of the main tropical climates summarized in the following statistics.

	Temperature		*Rainfall*	
	Max.	Min.	Annual	Seasonal Distribution
A	31·1°C	15·6°C	1030 mm (40·6″)	900 mm (35″) in summer months
B	27·8°C	20·6°C	968 mm (38·1″)	slight summer maximum
C	21·1°C	13·9°C	810 mm (31·9″)	500 mm (20″) in summer months
D	27·8°C	26·1°C	2250 mm (88·7″)	no dry season

11 What is shifting agriculture, and why is it common in tropical rain forests?

12 Why are the tropics not important for supplying meat and dairy products to the world market?

13 Why are some desert areas sparsely populated, while others are very densely populated?

Warm Temperate Climates

13

General Features

All temperate climates have a well-defined summer and winter as a result of the marked variation in angle of the sun's rays in the course of a year; but the warm temperate climates are noticeably warmer than the cool temperate climates.

The warm temperate region extends for about 10° on either side of 35° latitude, but its extent varies with local conditions. It lies between the trade winds which border the tropics and the westerlies of cool temperate latitude (fig. 108) being affected by the former in summer and the latter in winter. This seasonal change is due to the migration of the pressure belts with the overhead sun. It results in the summers having much more dependable weather than the winters.

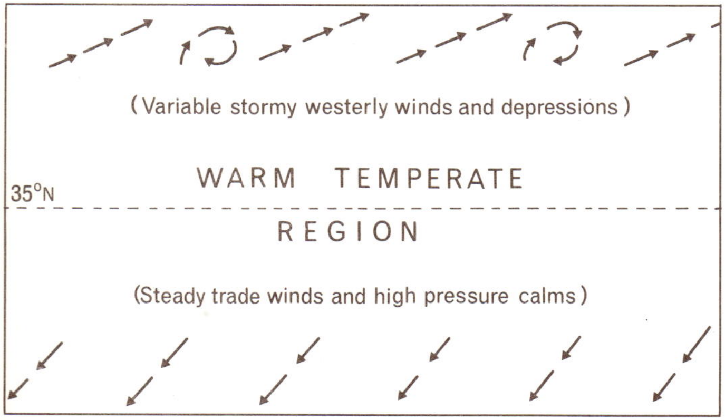

Fig. 108 Location of warm temperature region between two air masses

There are slight temperature variations within the region, depending on whether a place is at the east or west coast, or far inland, but it is rainfall differences which justify the subdivision into the following three climate types:

6 Warm temperate west margin (Mediterranean) with winter rain and summer drought.

7 Warm temperate east margin with rain all year but a summer maximum.

8 Warm temperate interior with summer rain and drought in winter.

6 Warm Temperate West Margin (Mediterranean) Climate

LOCATION

As its name suggests, this type of climate is found on coastal lowlands around the Mediterranean sea and on west coasts of continents in warm temperate latitudes (fig. 109). The main areas therefore are:

Mediterranean coastlands
Central California
Central Chile
Cape Province of South Africa

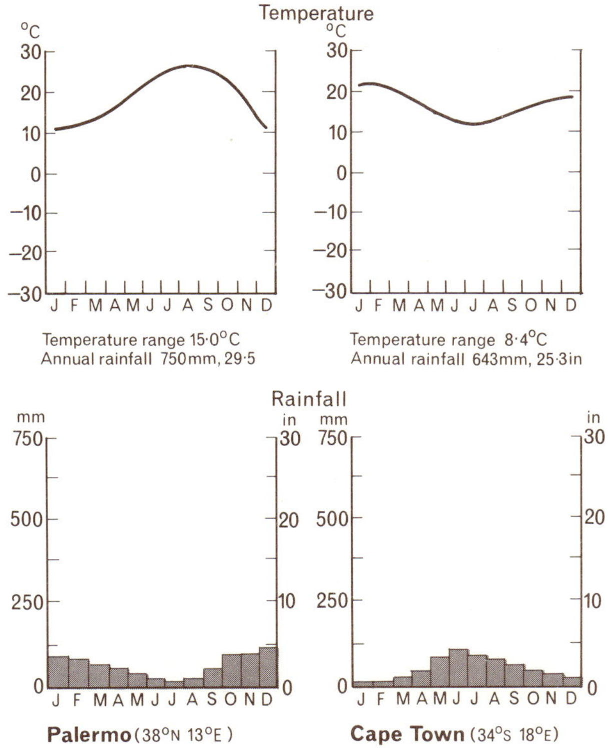

Fig. 109 Two 'Mediterranean' stations

The south-west tips of Australia around Perth and Adelaide.

FACTORS

In winter these areas are affected by wet westerly winds and depressions. Temperatures are kept fairly high, since on the poleward side there are oceans in the southern hemisphere and mountain barriers to keep out cold continental air in the northern hemisphere. In summer, the stable air of the mid-latitude anticyclone gives much calmer summer and drier weather.

TEMPERATURE

Frost is rare in these areas although it can occur, especially when cold continental air enters the region in winter. Snow does occasionally fall on the French Riviera and the Algerian coast. In normal years cultivation can continue throughout the winter, and tender fruits like olives and oranges are commonly grown. 'Out of season' vegetables are grown for the industrial north-west of Europe in countries such as Spain and Israel, and the mild winters result in a large 'off season' tourist trade.

The summers are almost as hot as in many parts of the tropics, often rising to about 25°C, and diurnal ranges are a little greater than in winter.

RAINFALL

Rainfall is almost entirely of the cyclonic type, although there are occasional convectional downpours, particularly in spring and autumn when the land is hot and the air tends to be unstable. This is the only type of climate (apart from deserts) with summer drought. While this poses difficulties for the farmer, it results in the 'Mediterranean' areas being very popular with holidaymakers. Annual rainfall totals are not great, seldom rising over 760 mm (30 inches). There is a noticeable decrease towards the tropics, with a gradual change to semi-desert conditions.

NATURAL VEGETATION

Mediterranean woodland is found in areas with a Mediterranean type of climate. It is specially adapted to surviving the summer drought, followed by a period of slow growth during the winter. In the wetter areas evergreen cork oak, cedar, and cypress trees are common but, where the rainfall is lighter, a scrub of bushes such as broom, gorse and myrtle is common. This scrub vegetation is called **maquis** where it is dense; but on limestone, where there is less surface water available and the plants are more widely dispersed, it is known as **garigue**.

WAYS OF LIFE

The 'Mediterranean' areas are generally very attractive for human settlement, as they have neither the excessive heat of tropical latitudes nor the unpleasantly cold winters of the cool temperate areas. Around the shores of the Mediterranean sea itself, the early Greek and Roman civilizations developed. In the 'new countries', similar areas such as California, Cape Province, and Victoria have been among the most popular with settlers. For people living in northern Europe, the French Riviera, Spain, Italy, and the Greek islands with their abundant sunshine have for long been favourite holiday areas. In recent years the annual summer migration has increased to such an extent that its influence is now being felt along the North African coast. The mildness of the winters results in a very long tourist season, so that resorts in this area do not have the problem of having large hotels and tourist attractions virtually unused for over half the year, as happens in many British holiday areas. The growth of the tourist business has had an important impact on the economy of the region, and in some parts it is the main source of employment and wealth.

Inland, undisturbed by this coastal tourist trade, farming continues much as it has done since Roman times. Because of the summer drought, grass does not grow well, and cattle rearing is almost unknown; on the rocky hillsides, sheep and goats graze on the coarse dry vegetation. The goats provide a certain amount of milk, but olive oil is the main substitute for animal fats. In the hill areas population is sparse, and people are generally poor; but on the coastal plains there are numerous farming villages where the people cultivate the land very intensively. Most of the fields are small, and irrigation is practised wherever possible. Most of the land is cultivated for fruit and vegetables; a two-tier system is often used with melons, tomatoes and lettuce growing under olive and orange trees. Where the fields are large enough, wheat is grown during the winter. In this way grain, vegetables, and fruit trees are harvested from one piece of land. The slopes bordering the plains are frequently terraced for vines to provide the wine which is almost essential in this area of water shortage. Although much of the produce is used locally, early vegetables like potatoes, tomatoes, citrus fruits, and wine are valuable exports and form the basis of the economy for most of the Mediterranean.

In the 'Mediterranean' areas of Africa, Australia, and America, fruit and vegetable growing is also very important. Since there is less pressure on the land and there is no native peasant population, the farms are larger and more highly mechanized. In California, large areas are given over to the commercial cultivation of cotton, and some of the orchards are almost big enough to be called plantations.

7a Warm Temperate East Margin Climate

LOCATION

The warm temperate climate is found on the east side of continents in warm temperate latitudes (fig. 110). The main areas are:

- South-east United States
- Lowlands around the River Plate
- Coastal New South Wales
- Natal
- Southern Japan
- North Island, New Zealand

FACTORS

In summer the wind tends to be on shore and gives a

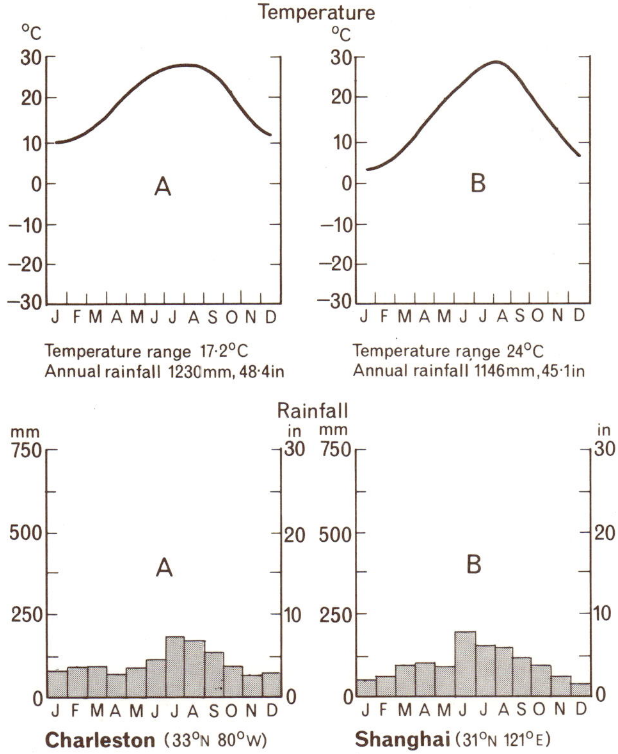

Fig. 110 *A* **Warm temperate east coast station**
B **Temperate monsoon station**

considerable amount of cloud and rain, especially if there are mountains near the coast. These conditions move into the tropics in winter and are replaced by westerly air which contains cyclonic disturbances. Although this westerly air has crossed a considerable amount of land, it does collect some moisture on approaching the coast and gives winter rain.

TEMPERATURE

Temperature is very similar to that found in Mediterranean areas, although there is a slightly greater chance of frost in winter.

RAINFALL

Rainfall varies greatly from place to place, but there is no dry season and summer tends to be wetter than winter.

7b Temperate Monsoon or China Type Climate

LOCATION

The temperate monsoon climate is confined to the east coast of China in latitudes which have climate type 7a on other east coasts (fig. 110).

FACTORS

In summer, onshore winds give conditions very similar to those found on other east coasts in these latitudes, but in winter the high pressure over Asia results in dry cold winds blowing from the interior of the continent. This reversal of wind direction is similar to that found in tropical monsoon areas.

TEMPERATURE

Summer temperatures are high, as in other warm temperate areas, but winters are cold with averages below 5°C and frequent frost. This cold winter is one of the main distinguishing features of the climate.

RAINFALL

Average annual rainfall varies from over 1500 mm (60 inches) in the south to about 500 mm (20 inches) in the north, with no dry season but a clear summer maximum.

NATURAL VEGETATION

The warm temperate east coasts all have forest

vegetation, most of which is of the hardwood deciduous type. Pine trees grow on the sandy areas. In the Chinese monsoon districts bamboo, a type of grass, is one of the most valuable plants.

WAYS OF LIFE

Throughout the world much of the natural vegetation produced by the climate has been cleared to permit large-scale farming. Although the style of farming varies from the plantations and truck farms of the United States to the communes of China, the products are similar everywhere. Cotton, maize, fruit, and wheat are grown along with many other crops, and cattle are reared for both meat and milk. Much of the farming is commercial, and these areas are generally prosperous and well populated.

8 Warm Temperate Interior Climate

LOCATION

The areas with warm temperate climate are inland from the east margin areas (fig. 111) and the main places where it is found are:

Interior plains of the Mississippi basin
Murray/Darling plains of Australia
Plains inland from the River Plate estuary
Plateau of South Africa
Plains of Central China

FACTORS

Winds from the sea bring humid air to these areas in summer. Temperature is high, pressure low, and the air unstable, with considerable rainfall (often of the convectional type) resulting. In winter the air movement is generally from the land, and rainfall is much lighter, especially at places far from the sea.

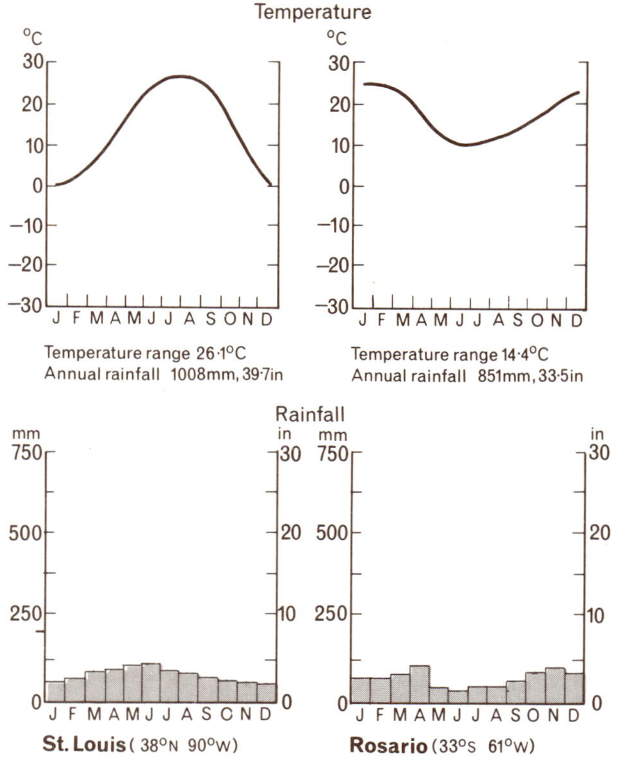

Fig. 111 Two warm temperate interior stations Note greater range of temperature in northern hemisphere.

TEMPERATURE

Summer temperatures are high and similar to those found in other warm temperate areas. In winter, the southern hemisphere areas have temperatures only slightly lower than those of coastal areas. In the northern hemisphere, where land masses are much more extensive, it becomes cold in winter with frost and some snow.

RAINFALL

Annual rainfall totals are rather lower than those

found in coastal areas, and there is a steady decrease inland. Everywhere, however, there is a clear summer maximum with a fairly dry winter.

NATURAL VEGETATION

Rainfall is insufficient to support forest vegetation and, although some trees do grow in the wetter parts, grass is very widespread. The grasses are not so tall or coarse as those growing in the tropical savannas. Although in many places it is tufted, turf-forming varieties are also common. In the northern hemisphere the grasslands are very extensive and continue into the cool temperate continental areas (Chapter 14), but in the southern hemisphere they are restricted by the narrowing of the continents.

WAYS OF LIFE

These are the great cattle-rearing areas of the world, and on the **pampas** of South America the animals are reared on large estates called *estancias*. The development of commercial beef-cattle farming coincided with the introduction of chilling and canning as a means of preserving the meat for transport to European markets. Grain crops also grow well on ploughed grasslands, and wheat and maize are very widespread. In Australia much of the wheat is exported, while in the United States and South Africa maize is often used for fattening beef cattle. Because of the large size of the farms, population density is not great, but most of the farms are prosperous and the standard of living is high.

STUDY SECTION ON CHAPTER 12

1 Which air masses affect warm temperate latitudes
(*a*) in summer,
(*b*) in winter?

2 Which warm temperate climates have
(*a*) summer drought,
(*b*) winter drought?

3 Name three areas, each in a different continent and far from the Mediterranean Sea, which have a Mediterranean climate.

4 Why do the temperate monsoon areas have colder winters than the other east-coast areas?

5 Which warm temperate climates are summarized below?

	Temperature		*Rainfall*	
	July	January	Annual	Seasonal Distribution
A	18·9°C	6·7°C	573 mm (22·6″)	480 mm (19″) September–April
B	28·3°C	6·1°C	1300 mm (51·2″)	all year
C	12·8°C	21·1°C	643 mm (25·3″)	530 mm (21″) April–October

6 Which of the warm temperate climates is important for the production of (*a*) cotton, (*b*) beef, (*c*) oranges? In each case name one country from which the product is exported.

Cool Temperate and Cold Climates

14

General Features

Cool temperate and cold climates are found in the parts of the world where the air movement is from west to east. In the lower atmosphere this general westerly air flow is marred by:

(*a*) depressions which give very variable weather,

(*b*) the development of large high-pressure areas over Asia, and North America in winter. These high-pressure areas often divert depressions south into the warm temperate region and keep temperate damp oceanic air out of the continental interiors. Climates of this type have a great variation in temperature and rainfall, which result in four main subdivisions being clearly recognized.

9 Cool Temperate West Margin or Maritime climate with a small temperature range and no dry season.

10 Cool Temperate Interior or Continental climate with a very large temperature range, and light rainfall which normally has a summer maximum.

11 Cool Temperate Desert climate, which also has a large temperature range but very light rainfall.

12 Cool Temperate East Margin climate, which has a temperature range larger than Maritime and smaller than Continental. Rainfall is fairly heavy and there is no dry season.

13 Polar or mountain climate.

9 Cool Temperate West Margin (Maritime) Climate

LOCATION

As the name suggests this type is found on the west side of a continent in cool temperate latitudes (fig. 112A). The main areas are:

North-west Europe (including the British Isles)
British Columbia and southern Alaska
Southern Chile
South Island of New Zealand.

FACTORS

All areas with this type of climate are dominated by:

1 Oceanic air which is humid and has a small annual temperature range

2 Depressions giving very variable weather with no easily recognized patterns

3 Warm ocean currents which make the on-shore winds unusually mild for the latitude.

TEMPERATURE

Summers are not very warm. Temperatures seldom

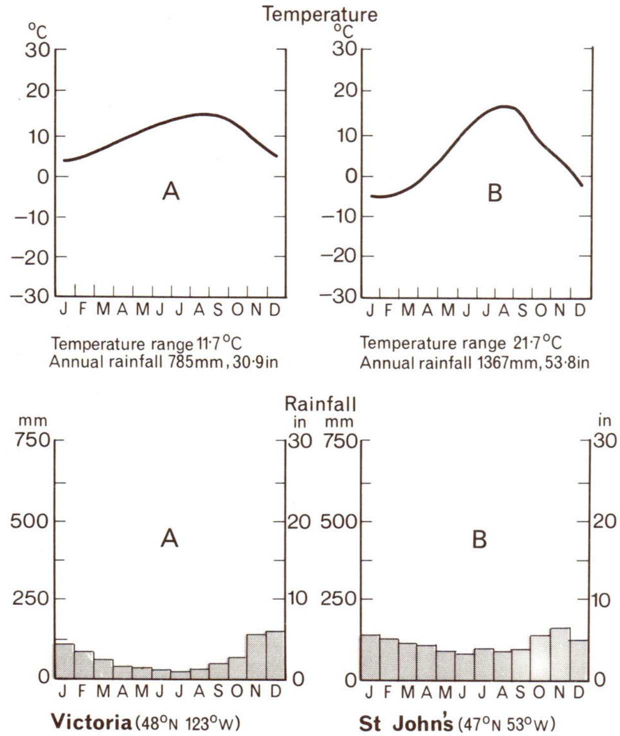

Fig. 112 *A* **Cool temperate west margin**
B **Cool temperate east margin**

rise as high as 20°C in the warmest month, but winters are extremely mild for places so far from the equator. Although frost and snow are common, the sea never freezes and the average for the coldest month is usually above 0°C.

RAINFALL

The predominance of humid oceanic air results in moderate or high rainfall totals except in well-formed rain shadows. Since much of the rain is of the cyclonic type, it is usually widespread when it does occur. There is no dry season.

NATURAL VEGETATION

The maritime areas, with their abundant rainfall throughout the year and equable temperatures, all have a natural vegetation of mixed forest in which both coniferous and deciduous trees are present. Where deciduous trees like oak, beech, and sycamore predominate, as in the natural vegetation of western Europe, the forests are leafless with little undergrowth during the winter. In summer the branches are clothed with leaves almost to ground level, and there is dense undergrowth. In North America coniferous trees are much more common, and they include the giant Douglas fir. In Europe much of the natural vegetation has been removed to provide land for farming, but in the Americas and Australasia much of the land is still clothed in its blanket of trees.

WAYS OF LIFE

It is very difficult to generalize about the ways in which the maritime areas have been developed, as they vary from the intensively farmed and highly industrialized countries of Western Europe to the almost uninhabited forests of southern Chile. In all areas where the land has been cleared for farming, however, the soil has been found to be moderately fertile and the climate well suited to mixed farming. The temperate grains of wheat, barley and oats are grown along with root crops like turnips and potatoes, and numerous sheep, beef cattle and dairy cows are reared. Even without industry the maritime areas can support a fairly large population, and there are almost certainly large parts of British Columbia, Washington and Southern Chile which could support many more people than they do at present.

Forestry is quite important in British Columbia, but southern Chile is too remote from the major industrial countries to be a major timber producer.

10 Cool Temperate Interior (Continental) Climate

LOCATION

Since there are no land masses in the cool temperate parts of the southern hemisphere, this type of climate does not occur south of the equator (fig. 113). It is found in:

the interior of North America (northern United States and Canada)
eastern Europe
west and central Asia

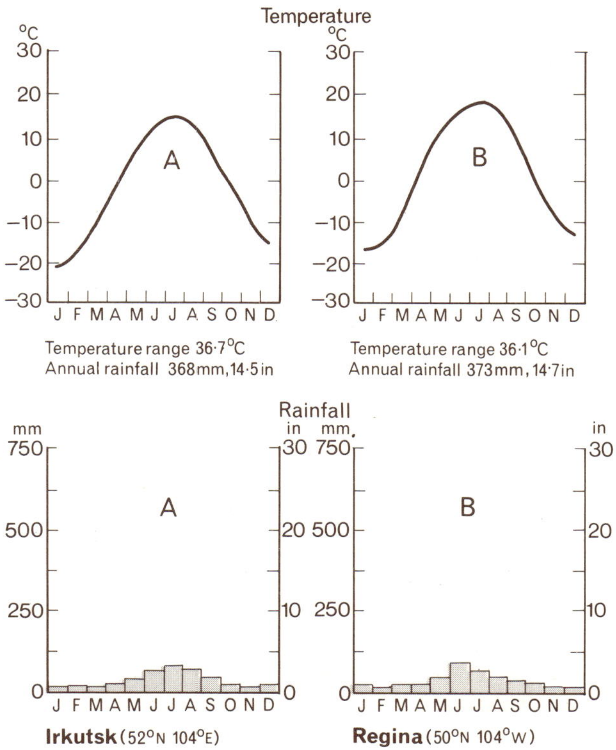

Fig. 113 Two cool temperate interior stations Note similarity in climate between two places thousands of miles apart.

FACTORS

At all times of the year remoteness from the sea is an important factor, but it is most noticeable in winter when high pressure keeps out almost all marine influence. In summer low pressure allows a little sea air to penetrate to these areas.

TEMPERATURE

The largest annual temperature ranges to occur anywhere in the world are found in these areas. This is mainly brought about by the very low winter temperatures. Winter averages of -10°C are common in North America, but the larger land mass of Asia produces averages of -20°C, while Verkhoyansk has an average January temperature of -50°C. Although the winters are very cold, the weather is not unpleasant since, during the short days, the winds are light, skies clear, humidity is low, and there is considerable sunshine. The worst weather occurs when depressions occasionally penetrate the high pressure to give gales and blizzards.

Although summers are short, the days are long and sunny with the average July temperature rising to 16°C.

RAINFALL

Rainfall is light since, even when sea air penetrates this distance inland, it has lost much of its moisture. Although annual totals are often under 625 mm (25 inches), this is ample for plant growth since most of it falls in summer. Any winter precipitation is in the form of snow and becomes available to plants when it melts in the spring.

NATURAL VEGETATION

The southern parts of these continental areas are too dry for trees, because of the high evaporation rate during the hot summers. The steppes of Russia like the Canadian prairies are northward continuations of the warm temperate grasslands. Further north, where the summers are shorter and cooler, lie the great coniferous forests which, along with the tropical rain forests, make up the main natural woodlands remaining in the world today.

Coniferous trees are specially adapted to survive a long cold winter, and they normally replace the deciduous hardwood forest where the growing season falls to about six months. They can also grow in drier conditions than the deciduous trees, and therefore replace them in the interior of continents and on dry sandy soils. Their greatest extent is in the northern part of North America and Eurasia where they extend the full width of the continents. The trees are thin tall and straight, with only a small number of species, and little undergrowth to hinder lumbering operations. As a result these forests are the principal ones to be exploited at present, and they provide much valuable raw material for the building and paper industry. Fir, pine, and spruce are common varieties, but towards the Arctic Circle the conifers give way to the deciduous birch and larch which can tolerate a very short growing season.

WAYS OF LIFE

In many ways the development of the steppes and the prairies resembles that of the warm temperate grasslands. The longer colder winters make farming a little more difficult, and they are not so densely populated. They are the main wheat-growing areas of the world and much of the farming is highly mechanized. The population density decreases steadily northwards, and the forest regions are very sparsely populated. A small number of people are engaged in the trapping of fur-bearing animals like mink and fox, but the main work is connected with forestry. It is mostly along the southern edge of the forests where timber is readily accessible that the industry has developed. Although most of the world's soft wood comes from these areas, there are still vast tracts of untapped virgin forest.

11 Cool Temperate Desert Climate

LOCATION

In certain cool temperate interior areas rainfall is insufficient for much plant growth, and desert conditions result. The main area with this type of climate is in the U.S.S.R. east of the Caspian Sea, Mongolia, and western China around the Tarim

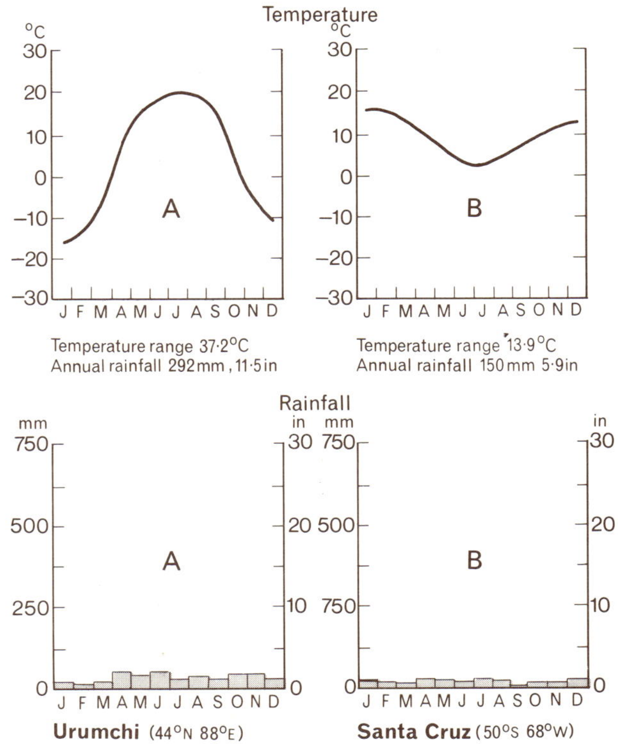

Fig. 114 Cool temperate deserts
A **Central Asia**
B **Patagonia**

Basin (fig. 114). A modified form of this climate occurs in the Patagonian region of Argentina.

FACTORS

The drought of these areas results from the very low humidity of the atmosphere brought about by remoteness from the sea and the rain shadow effect of very high mountains such as the Himalayas, Kunlun Shan, and the Andes.

TEMPERATURE

In Asia there is very little temperature difference between the desert areas and the slightly wetter cool temperate interior climate, but in Patagonia the marine influence gives temperatures similar to those of the maritime areas.

RAINFALL

In all these areas rainfall is very slight, and many places are even drier than some parts of the tropical deserts.

NATURAL VEGETATION

These deserts do not have such extensive areas of sand and bare rock as the tropical deserts, because the lower temperatures, especially in winter, greatly reduce the evaporation rate. A little coarse tussocky grass can grow in most parts.

WAYS OF LIFE

Pastoralism is universal except in a few places where spring water permits a little cultivation. Because of the very thin nature of the vegetation, the carrying capacity of the land is very low, and nomadism is the traditional way of life in Central Asia. Tribes such as the Kirghiz keep sheep, goats, and horses. In much the same way as the Bedouin of the Sahara, they migrate from one area of poor pasture to another, always avoiding damage by overgrazing. Patagonia has an area of commercial sheep farming but, because of the low carrying capacity, farms have to be very large. In North America there are no deserts of this type, but the High Plain to the east of the Rockies has a very low rainfall. Unless irrigation water is available, the land is used for extensive cattle ranching.

12 Cool Temperate East Margin Climate

LOCATION

Cool temperate east margin climate occurs on the opposite coasts of continents from the maritime areas. Once again it is confined to the northern hemisphere, although, as has already been seen, Patagonia has temperatures of the type occurring in this climate (fig. 114B). The two principal regions are:

Eastern U.S.S.R.
Eastern Canada.

FACTORS

In many ways these areas are transitional between the maritime and continental climates. The general air movement is again from the west, bringing temperature extremes and drought of the continental areas, but proximity to the sea moderates temperatures and increases rainfall.

TEMPERATURE

Summer temperatures are fairly high, rising to near 20°C. Winters are colder than in maritime areas, because of the air movement from the continental interior and the presence of cold ocean currents. Averages of around 0°C or a little lower are quite common.

RAINFALL

Rainfall is mostly of cyclonic origin. The disturbances which produce it collect moist sea air as they approach the coast. Annual rainfall totals are similar to those of maritime areas, being normally over 750 mm (30 inches) with no dry season.

NATURAL VEGETATION

Like the maritime areas these are lands of mixed forests. Here coniferous trees predominate over the deciduous ones, and in many ways they resemble the great northern forests of Canada and Russia.

WAYS OF LIFE

In eastern Canada much of the more accessible timber has been cut. Now second-growth forests are producing wood for the large wood pulp and paper mills of the Maritime Provinces and St. Lawrence lowlands. In areas where the climate and soil are favourable, however, the trees have been completely removed. Mixed farming, similar to that of western Europe, is practised to supply the nearby industrial towns with fresh milk, meat and vegetables.

The winters of eastern Siberia are much more severe than those of eastern Canada, and the area is very sparsely populated. In Manchuria to the south, where the forest was thinner and the weather warmer, the land has been cleared in places for cultivation. Wheat, millet, and soya beans are grown.

13 Polar Climate

FACTORS

The areas inside the Arctic and Antarctic Circles have high pressure and are not affected by the depressions of cool temperate latitudes (fig. 115B).

In summer, when there is continual daylight, the sun feels quite warm, but air temperatures seldom rise much above freezing-point due to the presence of extensive areas of snow and ice. Along the arctic coasts of Canada and the U.S.S.R. the average for July may be as high as 10°C. In the Antarctic, where the ice-covered continent has a greater chilling effect than the frozen Arctic Ocean, the average for the hottest month is about −5°C.

Summers are very short, lasting for about 6 weeks, and the low altitude of the sun produces little heat in spring and autumn.

In winter, when there is continual darkness, temperatures drop very low, giving an average of −40°C for some parts of the Antarctic coast. The actual temperatures are often well below the average and −127°C has been recorded in Antarctica.

Precipitation in these polar areas is very light (about 300 mm, 12 inches per year) because of the high pressure and stable air. When it does occur, it falls as snow, which builds up to great thicknesses over the years. In Antarctica and Greenland the snow has compacted itself into great icecaps, from which glaciers slowly slide to the sea before breaking off to form icebergs. A characteristic of these icecap regions is the strong winds which blow for most of the year from the interior towards the coast.

NATURAL VEGETATION

In northern latitudes, near and beyond the Arctic Circle, the growing season is too short for trees. They are replaced by perennial flowering herbs, heaths, and mosses. These plants lie dormant for most of the year, often protected from the cold and the strong winds of winter by a blanket of snow, ready to produce a colourful and varied vegetation of short plants in the summer. An important feature is the presence of permanently frozen subsoil (called **permafrost**) which leads to very marshy conditions in spring when the melt-water from the snow cannot drain into the ground.

There are no tundra trees of any significance in the southern hemisphere, as the latitudes in which they would occur are occupied by the great Southern Ocean; the ice cap of Antarctica prohibits any plant growth.

WAYS OF LIFE

As in all deserts, where there is little vegetation, a nomadic way of life is traditional in the tundra areas. In North America the Eskimos have lived for centuries by hunting animals such as the caribou and the seals which live on the shores of the Arctic Ocean. Many Eskimos are abandoning their nomadic existence now, and finding work in the mining camps which have recently grown up after the discovery of oil and other minerals in the frozen wastes of the north.

The pattern of life in Eurasia is similar. Here the domesticated reindeer has provided a more reliable source of food and, although the Lapps are nomadic, they are not so dependent on hunting and trapping as are the Eskimos.

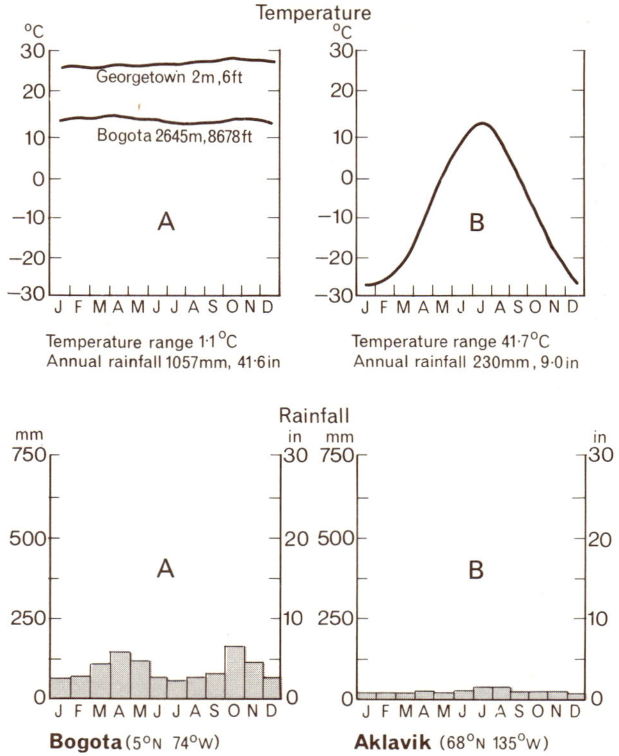

Fig. 115 ***A* Mountain climate at equator (Georgetown is given for comparison) *B* Tundra climate**

Mountain Climate

FACTORS

The conditions described for the climate regions numbered 1 to 13 are for lowland or low plateau areas. Many of these regions also contain mountain ranges where conditions differ considerably from those of the neighbouring lowlands (fig. 115A). The main factors which bring about these differences are as follows:

1. The decrease in temperature with altitude.
   This results in zones of progressively lower temperatures occurring one above the other on a mountain side. In many ways these are similar to the temperature zones which lie between the equator and the poles. The top zone is the region of permanent snow. The snow line which marks its lower limit falls from about 4500 metres (15,000 feet) in the tropics to 2500 metres (8000 feet) in cool temperate latitudes.
2. Ascending air currents caused when winds have to rise over mountains.
   This results in precipitation being much greater over mountain areas. Above about 2000 metres (7000 feet), the humidity of the air decreases with a resulting decrease in precipitation at very high altitudes.

3 Small proportion of land at high altitudes.
Thus there is little friction to slow down air movement and very strong winds are common.

4 Great variations in height and slope over a relatively short distance.
This results in much more frequent changes in rainfall, sunshine, and wind than are found in a lowland area of similar size.

Large high-level plateaus give extensive areas of climate which is very much cooler than that expected for the latitude. In parts of the tropics, such as East Africa, this has provided conditions which are attractive for European settlement; but in higher-latitude areas, such as Norway, the advantages of the mild maritime climate are lost on the cold windy fjeld.

One feature of the neighbouring lowland climate which persists to high levels is the annual temperature range. Tropical areas 3000 metres (10,000 feet) above sea-level have the average annual temperatures of cool temperate latitudes but also the small annual range of the tropics.

NATURAL VEGETATION

Fig. 116 shows the vegetation zones which are found on mountain slopes at different latitudes. The general pattern is for forest on the lower slopes to thin out gradually and give way to an area of grass, mosses, and small flowering plants which are collectively known as alpine pasture. In poorly drained areas the

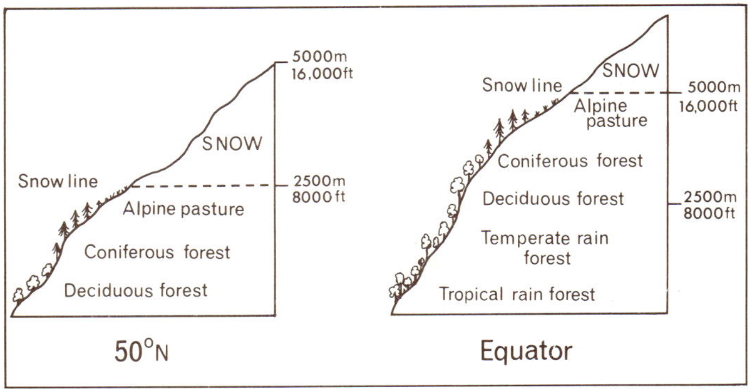

Fig. 116 Changes in vegetation with altitude where there is abundant rainfall

Fig. 117 Location of climate stations for which graphs are drawn in figs. 109–115

alpine pasture is replaced by coarse plants of the heath, gorse, and bracken type commonly found on British moorlands. In most mountain areas some snow falls in the winter but, as already mentioned, if they are high enough there is a region of permanent snow above the alpine pasture.

WAYS OF LIFE

Mountain areas, because of their high degree of relief, thin soil, cool damp climate, and difficult communications, tend to be sparsely populated. Forestry is a common occupation and in many places the timber provides the building material for houses, and fuel for the fires, as well as the principal article for sale. A certain amount of pastoralism is also common. Often it is of the transhumance type where the animals are kept on the alpine pasture in summer, while clearings on lower ground are used to grow fodder crops for winter use.

STUDY SECTION FOR CHAPTER 14

1 State two features of the atmosphere which spoil the general air movement from west to east in parts of the cool temperate region.

2 Which cool temperate climate has:
(*a*) the smallest annual temperature range,
(*b*) summer rain and winter drought,
(*c*) the largest annual temperature range?

3 Why is the cool temperate west coast climate sometimes called *maritime*?

4 Why is it that the cold *continental* winters are sometimes quite pleasant?

5 Why is an annual rainfall of 380 mm (15 inches) sufficient for unirrigated cultivation in continental areas but not in tropical areas?

6 How does the climate of Halifax (Nova Scotia) differ from that of Victoria (B.C.)?

7 List two ways in which the climate of a mountain area differs from that of neighbouring lowlands, and one way in which it is similar.

8 Identify the cool temperate climate summarized in the following statistics. In each case explain why it has the characteristics which the statistics show.

	Temperature		*Rainfall*	
	January	July	Annual	Seasonal Distribution
A	1·1°C	14·4°C	2142 mm (84·3″)	no dry season
B	−11·1°C	18·9°C	533 mm (21·0″)	most in summer

9 (This exercise refers to the whole of Section C.)
For each of the places A–H listed on the following table:

(*a*) Describe the climate—give
- (i) maximum temperature
- (ii) minimum temperature
- (iii) annual temperature range
- (iv) annual rainfall
- (v) seasonal distribution of rainfall.

(*b*) State which hemisphere it is in.
(*c*) Suggest which temperature region it is in.
(*d*) Suggest in which subdivision (west coast, east coast, interior) of the temperature region it is located.
(*e*) Name the type of climate.
(*f*) State the areas in the world where the place might be located.
(*g*) Name and describe briefly its natural vegetation.

		J	F	M	A	M	J	J	A	S	O	N	D
A	Temp. °C.	1·7	1·7	2·8	7·2	11·1	14·4	16·7	15·6	12·7	8·3	5·0	2·2
	Rain mm	216	163	150	104	115	97	147	190	221	226	211	216
	in	8·5	6·4	5·9	4·1	4·5	3·8	5·8	7·5	8·7	8·9	8·3	8·5
B	Temp. °C	17·8	17·8	16·7	15·0	13·3	12·2	11·7	12·2	12·8	13·9	15·6	16·7
	Rain mm	0	0	13	3	69	152	135	86	10	13	8	0
	in	0	0	0·5	0·1	2·7	6·0	5·3	3·4	0·4	0·5	0·3	0
C	Temp. °C	−10·6	−8·3	−3·9	4·4	10·0	13·3	16·1	15·0	10·0	4·4	−2·8	−2·3
	Rain mm	13	13	18	18	59	74	66	64	33	15	18	13
	in	0·5	0·5	0·7	0·7	2·3	2·9	2·6	2·5	1·3	0·6	0·7	0·5
D	Temp °C	21·1	23·9	27·8	31·1	30·6	28·3	26·1	25·0	26·1	26·7	24·4	22·2
	Rain mm	0	0	3	13	81	122	175	287	127	10	0	0
	in	0	0	0·1	0·5	3·2	4·8	6·9	11·3	5·0	0·4	0	0
*E	Temp. °C	15·0	15·0	15·0	14·4	14·4	14·4	14·4	15·0	15·0	15·0	14·4	15·0
	Rain mm	107	109	132	188	127	38	23	38	76	94	97	97
	in	4·2	4·3	5·2	7·4	5·0	1·5	0·9	1·5	3·0	3·7	3·8	3·8
F	Temp. °C	23·9	23·9	26·1	28·3	30·0	28·3	27·2	26·7	26·7	28·3	27·2	25·6
	Rain mm	3	0	3	0	18	523	693	406	300	61	10	0
	in	0·1	0	0·1	0	0·7	20·6	27·3	16·0	11·8	2·4	0·4	0
G	Temp. °C	31·1	24·4	28·9	25·6	22·2	18·3	17·8	20·0	23·3	27·8	29·4	31·1
	Rain mm	130	124	69	23	10	8	13	3	13	13	28	76
	in	5·1	4·9	2·7	0·9	0·4	0·3	0·5	0·1	0·5	0·5	1·1	3·0
H	Temp. °C	25·5	26·1	25·0	23·3	21·7	21·1	20·6	21·1	21·1	21·7	22·8	24·4
	Rain mm	127	114	135	107	81	56	43	46	66	84	104	140
	in	5·0	4·5	5·3	4·2	3·2	2·2	1·7	1·8	2·6	3·3	4·1	5·5

* All near sea-level except E which is 2740 m (9000 feet) above sea-level.

I. Meteorological Office Services

Weather forecasts issued by the Meteorological Office, London Road, Bracknell, Berkshire are broadcast on radio and television at the times shown in the *Radio Times*. Map A shows the areas covered by the district names used in these forecasts.

Special local forecasts can be obtained from the following district Meteorological Offices by post or telephone (see Weather in the telephone directory).

England and Wales

Abingdon (Berks.), Acklington (Northumberland), Bassingbourn (Herts.), Bawtry (near Doncaster), Birmingham Airport, Boscombe Down (Wilts.), Chivenor (Devon), Glamorgan Airport, Gloucester, Linton upon Ouse (Yorks.), Liverpool Airport, London Weather Centre, Lyneham (Wilts.), Manby

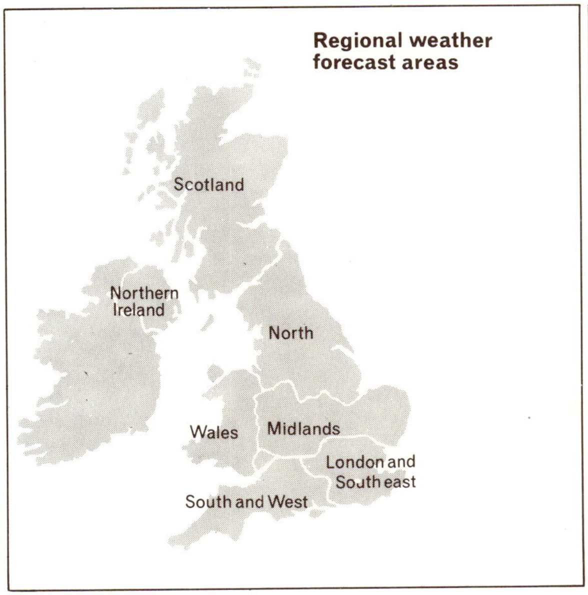

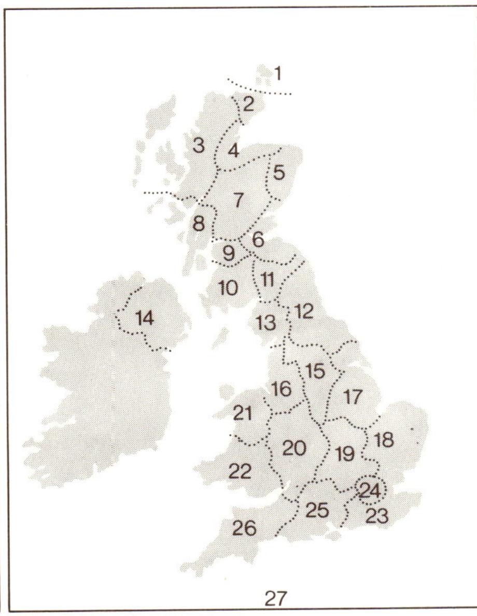

Forecast Districts

1. Orkney and Shetland
2. Caithness
3. North west Scotland
4. Moray Firth area
5. Aberdeen area
6. Edinburgh and East Scotland
7. Central Highlands
8. Argyll
9. Glasgow area
10. South west Scotland
11. Borders
12. North east England
13. Lake District
14. Northern Ireland
15. Central Northern England
16. North west England
17. Eastern England
18. East Anglia
19. East Midlands
20. West Midlands
21. North Wales
22. South Wales
23. South east England
24. Greater London
25. Central Southern England
26. South west England
27. Channel Islands

Map A Weather Forecast Areas and Districts

(Lincs.), Manchester Weather Centre, Marham (Norfolk), Mildenhall (Suffolk), Newcastle Weather Centre, Oakington (near Cambridge), Plymouth (Devon), Preston (Lancs.), St. Mawgan (Cornwall), Shawbury (near Shrewsbury), Southampton Weather Centre, Thorney Island (Hants.), Upavon (Wilts.), Valley (Anglesey), Watnall (Nottingham), Wittering (Northants), Wyton (Hunts.).

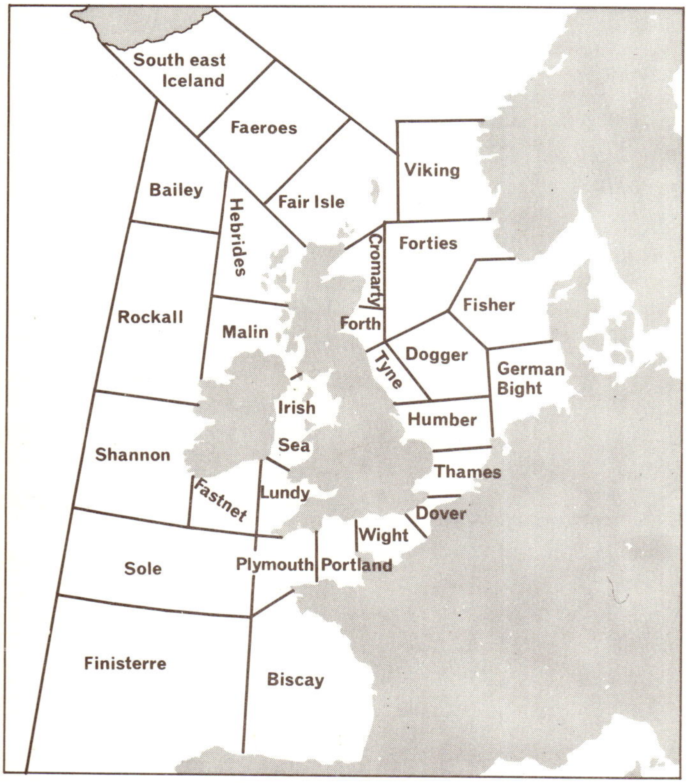

Map B Sea Areas for Weather Reports

Scotland

Aberdeen Airport, Edinburgh Airport, Glasgow Weather Centre, Kinloss (Morayshire), Kirkwall Airport (Orkney), Leuchars (Fife), Pitreavie (near Dunfermline), Prestwick Airport.

Weather Reports from coastal stations and forecasts for sea areas are broadcast separately from the general forecasts and Map B shows the locations of the various sea areas.

A Daily Weather Report is published giving the following information:

(*a*) A weather map for much of the northern hemisphere.
(*b*) Weather maps for the British Isles and adjacent areas for 1800 h on the previous day, 0000 h and 0600 h on the date of issue.
(*c*) A general statement of the atmospheric conditions shown on these maps.
(*d*) Summary of the past month's weather.

These daily weather reports can be obtained at a reduced subscription rate by schools and colleges.

II. Climate Statistics for Selected Places in the World

(The map on page 111 shows the location of the places.)

SCOTLAND

STATION	Altitude	Temp. °C Rain mm	J	F	M	A	M	J	J	A	S	O	N	D	Average Total
ABERDEEN	58 m	T	2·4	2·9	4·5	6·6	9·0	12·0	14·0	13·6	11·7	8·8	5·6	3·7	7·9°C
	(190′)	R	77	57	52	56	68	52	84	58	77	91	89	79	840 mm
BRAEMAR	339 m	T	0·6	1·0	2·8	5·2	8·3	11·4	13·1	12·6	10·2	7·0	3·8	2·0	6·5°C
	(1113′)	R	104	68	56	57	66	48	76	78	75	105	96	94	923 mm
ESKDALEMUIR	242 m	T	1·4	1·8	3·6	5·8	8·9	11·7	13·3	13·0	10·8	7·6	4·6	2·8	7·1°C
	(794′)	R	190	117	100	96	95	100	127	140	135	168	150	159	1577 mm
GREENOCK	61 m	T	3·8	4·1	5·8	8·0	11·0	13·7	15·1	14·8	12·8	9·8	6·9	5·2	9·2°C
	(200′)	R	199	129	103	90	89	85	103	110	136	188	167	177	1576 mm
INVERNESS	74 m	T	3·0	3·2	5·0	7·0	9·7	12·4	14·2	14·0	12·0	8·9	5·8	4·2	8·3°C
	(242′)	R	69	48	38	45	54	50	75	78	67	95	63	59	741 mm
KELSO	59 m	T	2·3	3·0	5·0	7·3	10·1	13·1	14·9	14·2	12·0	8·7	5·4	3·7	8·3°C
	(195′)	R	57	43	40	39	49	48	71	76	58	67	59	47	654 mm
LERWICK	82 m	T	3·1	2·9	3·9	5·5	7·8	10·0	12·0	12·1	10·6	8·1	5·9	4·4	7·2°C
	(269′)	R	115	84	78	69	56	53	64	70	95	113	118	113	1028 mm
NORTH BERWICK	40 m	T	3·3	3·8	5·4	7·6	9·9	12·8	14·8	14·5	12·8	9·5	6·5	4·7	8·8°C
	(130′)	R	54	37	41	37	54	49	68	78	62	67	59	46	652 mm
TIREE	9 m	T	5·3	5·2	6·3	7·8	10·1	12·2	13·7	13·8	12·6	10·4	7·8	6·4	9·3°C
	(29′)	R	122	79	68	66	53	71	89	91	114	142	124	119	1139 mm
WICK	36 m	T	3·4	3·5	4·8	6·4	8·4	10·8	12·9	12·7	11·5	9·0	6·4	4·6	7·9°C
	(119′)	R	74	50	46	51	46	51	65	67	73	79	80	75	757 mm

ENGLAND AND WALES

STATION	Altitude	Temp. °C Rain mm	J	F	M	A	M	J	J	A	S	O	N	D	Average Total
BIRMINGHAM	130 m	T	3·2	3·7	5·8	8·7	11·5	15·0	16·7	16·4	13·9	10·0	6·6	4·4	9·7°C
	(425′)	R	76	54	48	54	63	48	46	69	59	70	77	67	731 mm
BUXTON	307 m	T	1·9	2·0	4·0	6·5	9·4	12·6	14·3	14·0	11·8	8·4	5·3	3·3	7·8°C
	(1007′)	R	145	102	71	81	74	81	102	104	96	129	132	122	1239 mm
CAMBRIDGE	12 m	T	3·4	3·9	6·1	8·8	11·9	15·1	17·0	16·7	14·3	10·4	6·7	4·5	9·9°C
	(41′)	R	48	33	30	43	48	38	58	48	50	53	56	40	545 mm
CARDIFF	62 m	T	4·4	4·6	6·6	9·1	11·9	15·0	16·4	16·5	14·3	10·9	7·6	5·5	10·2°C
	(202′)	R	117	76	58	63	76	56	86	99	91	114	117	109	1063 mm
KESWICK	77 m	T	3·7	3·8	5·8	7·8	10·9	13·7	15·2	15·0	12·8	9·8	6·7	5·0	9·2°C
	(254′)	R	178	107	84	84	79	86	114	129	142	170	157	154	1484 mm
LIVERPOOL	60 m	T	4·0	4·2	6·0	8·3	11·3	14·1	15·7	15·6	13·7	10·3	7·2	5·3	9·6°C
	(198′)	R	68	48	38	40	55	51	71	79	66	76	76	63	731 mm
ROSS ON WYE	68 m	T	4·2	4·5	6·4	8·9	11·7	14·9	16·5	16·2	14·0	10·4	7·2	5·3	10·0°C
	(223′)	R	76	51	35	51	55	38	58	58	58	71	79	71	701 mm
SCILLY Is.	50 m	T	7·7	7·3	8·5	9·7	11·7	14·3	16·0	16·5	15·2	12·7	10·3	8·6	11·5°C
	(163′)	R	91	68	61	56	56	43	56	61	61	89	94	89	825 mm
SOUTHAMPTON	20 m	T	4·6	4·9	7·0	9·7	12·5	15·7	17·1	17·3	14·9	11·3	7·8	5·7	10·7°C
	(65′)	R	89	58	48	53	53	46	58	63	63	84	96	86	803 mm
TYNEMOUTH	33 m	T	4·0	4·4	5·7	7·7	9·7	12·9	15·1	14·9	13·2	10·4	7·3	5·4	9·2°C
	(108′)	R	61	43	41	43	53	46	79	79	58	63	66	58	690 mm
YORK	17 m	T	3·3	4·0	6·0	8·6	11·6	14·7	16·6	16·2	13·9	10·2	6·8	4·7	9·7°C
	(57′)	R	61	43	33	43	51	48	61	63	53	56	63	51	626 mm

IRELAND

STATION	Altitude	Temp. °C Rain mm	J	F	M	A	M	J	J	A	S	O	N	D	Average Total
ALDERGROVE	67 m	T	3·7	4·2	5·9	7·9	10·5	13·3	14·7	14·5	12·7	9·7	6·6	4·9	9·1°C
	(220′)	R	81	53	48	48	53	68	91	81	81	84	73	91	852 mm
VALENTIA	9 m	T	7·2	7·2	7·8	8·9	11·7	13·9	15·0	15·0	13·9	11·7	8·9	7·8	10·6°C
	(30′)	R	178	120	96	79	81	81	109	117	109	129	160	160	1419 mm

60°N
Lerwick
Wick
Inverness
Aberdeen
Braemar
Tiree
North Berwick
Greenock
Kelso
Eskdalemuir
Tynemouth
Aldergrove
Keswick
York
Liverpool
Buxton
Birmingham
Cambridge
Valentia
Ross on Wye
Cardiff
Southampton
50°N
Scilly Is.

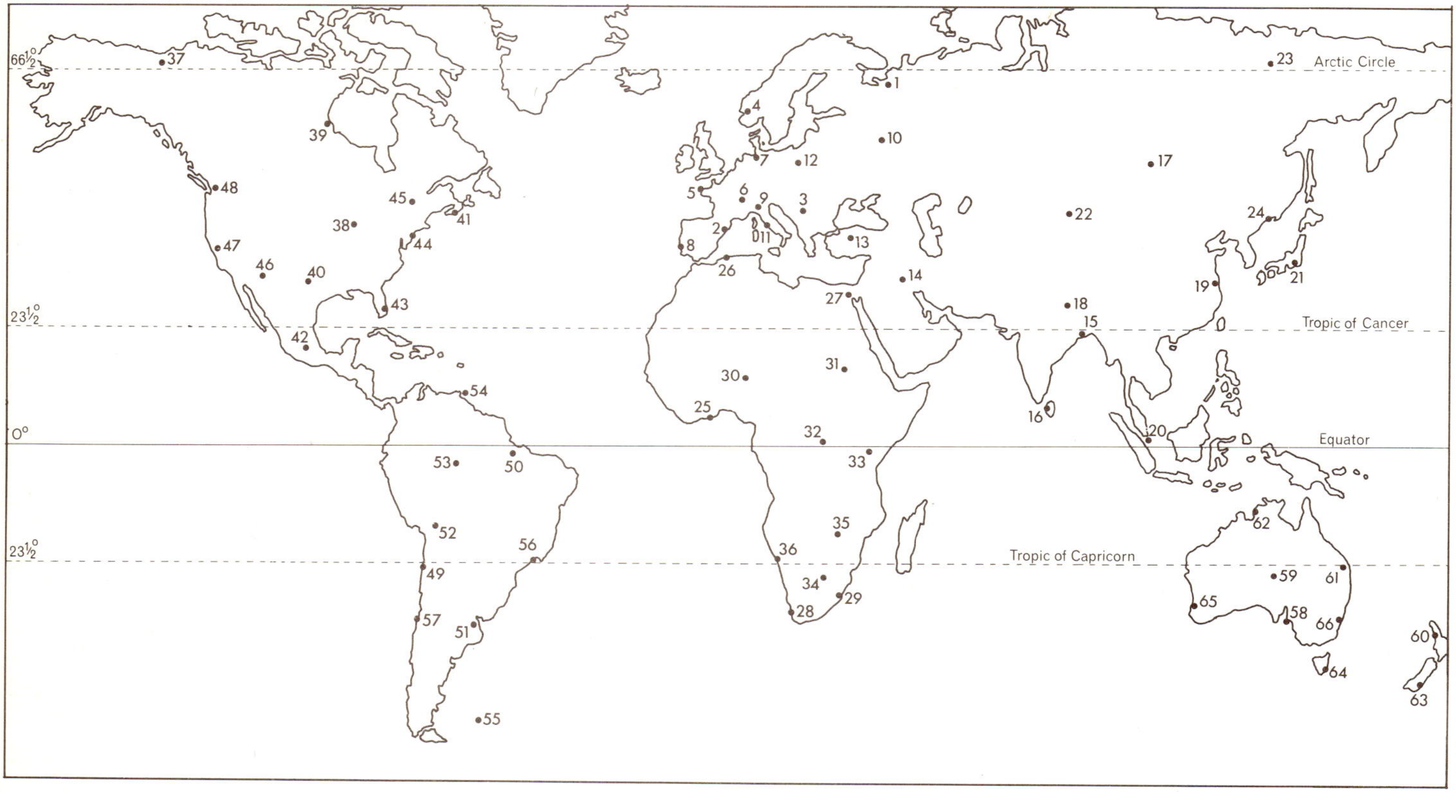

EUROPE

1. Archangel
2. Barcelona
3. Belgrade
4. Bergen
5. Brest
6. Geneva
7. Hamburg
8. Lisbon
9. Milan
10. Moscow
11. Rome
12. Warsaw

ASIA

13. Ankara
14. Baghdad
15. Calcutta
16. Colombo
17. Irkutsk
18. Katmandu
19. Shanghai
20. Singapore
21. Tokyo
22. Urumchi
23. Verkhoyansk
24. Vladivostok

AFRICA

25. Accra
26. Algiers
27. Cairo
28. Cape Town
29. Durban
30. Kano
31. Khartoum
32. Kisangani
33. Nairobi
34. Pretoria
35. Salisbury
36. Walvis Bay

NORTH AMERICA

37. Aklavik
38. Chicago
39. Churchill
40. Dallas
41. Halifax
42. Mexico City
43. Miami
44. New York
45. Ottawa
46. Phoenix
47. Sacramento
48. Vancouver

SOUTH AMERICA

49. Antofagasta
50. Belem
51. Buenos Aires
52. La Paz
53. Manaos
54. Port of Spain
55. Port Stanley
56. Rio de Janeiro
57. Valparaiso

AUSTRALASIA

58. Adelaide
59. Alice Springs
60. Auckland
61. Brisbane
62. Darwin
63. Dunedin
64. Hobart
65. Perth
66. Sydney

EUROPE

STATION	Altitude	Temp.°C Rain mm	J	F	M	A	M	J	J	A	S	O	N	D	Average Total
1 ARCHANGEL	7 m	T	–15·0	–14·4	–8·9	–1·1	4·4	10·6	13·9	12·2	7·2	0·6	–6·7	–12·2	–0·6°C
	(22′)	R	30·5	28 0	27·9	17·8	33·0	48·3	66·0	68·6	55·9	48·3	40·6	33·0	503·0 mm
2 BARCELONA	94m	T	8·9	10·0	12·1	13·8	17·9	21·2	23·9	23·8	22·3	17·8	13·3	10·0	16·7°C
	(310′)	R	30·5	53 3	48·2	45·8	45·7	33·0	30·5	43·2	66·0	83·5	68·5	45·8	597 mm
3 BELGRADE	137 m	T	0·0	1·1	6·7	12·8	17·2	20·0	22·8	22·3	18·2	13·3	7·2	1·7	11·7°C
	(450′)	R	40·6	33·0	40·5	56·0	66·0	71·1	48·3	63·5	43·2	68·6	45·7	48·3	624·8 mm
4 BERGEN	43 m	T	1·7	1·7	2·8	7·2	11·1	14·4	16·7	15·6	12·7	8·3	5·0	2·2	8·3°C
	(140′)	R	200·7	152·4	137·2	111·8	99·0	106·8	132·1	185·4	233·7	233·7	203·2	205·7	2001·5 mm
5 BREST	17 m	T	6·7	7·2	8·3	10·0	12·8	15·6	17·2	17·8	16·0	12·9	9·4	7·8	11·7°C
	(55′)	R	88·8	76·2	63·5	63·5	48·3	50·8	50·7	56·0	58·4	91·4	106·7	111·8	866·0 mm
6 GENEVA	405 m	T	1·0	2·3	6·1	10·0	13·9	17·8	19·4	18·9	15·6	10·6	5·6	1·7	10·0°C
	(1330′)	R	48·2	45·7	55·9	63·5	76·2	78·7	73·7	91·4	91·4	96·5	78·7	60·9	861·0 mm
7 HAMBURG	20 m	T	0·0	1·1	2·8	7·2	11·6	15·7	17·2	16·1	13·9	8·9	4·4	1·1	8·3°C
	(65′)	R	53·2	48 3	50·8	45·7	53·3	68·6	86·4	81·3	63·5	66·0	53·3	63·5	734·0 mm
8 LISBON	95 m	T	10·5	11·1	12·8	14·4	16·7	19·4	21·7	22·2	20·6	17·2	13·9	11·1	16·0°C
	(313′)	R	83·8	81·4	78·8	60·9	43·2	17·8	5·1	5·1	35·6	78·7	106·7	91·4	686 mm
9 MILAN	104 m	T	1·1	4·4	8·3	13·3	17·2	21·1	23·3	22·8	19·4	13·3	7·2	2·8	12·8°C
	(340′)	R	56	50 8	53·3	60·8	86·4	60·0	58·4	55·9	71·0	99·0	83·8	60·9	802·6 mm
10 MOSCOW	154 m	T	–9·4	–8·9	–4·4	3·9	12·2	16·7	18·3	16·6	11·2	4·4	–2·8	–7·8	3·9°C
	(505′)	R	38·2	35·6	27·9	48·3	55·9	73·7	76·2	73·7	48·3	68·6	43·2	40·6	630 mm
11 ROME	115 m	T	7·8	8·3	11·1	13·9	17·8	21·7	24·4	24·3	22·3	17·2	12·2	8·9	16·1°C
	(378′)	R	68·6	58·4	38·1	43·2	50·8	25·5	15·2	23·0	68·6	94·0	96·5	71·1	653·0 mm
12 WARSAW	120 m	T	–3·9	–2·8	1·1	7·8	11·1	16·7	18·3	17·8	11·1	8·3	2·2	–2·2	7·8°C
	(395′)	R	30·5	28·0	33·0	38·2	48·3	66·0	76·2	76·2	48·3	43·2	35·6	35·6	559 mm

ASIA

STATION	Altitude	Temp.°C Rain mm	J	F	M	A	M	J	J	A	S	O	N	D	Average Total
13 ANKARA	861 m	T	0·0	1·1	5·0	11·1	16·1	18·3	22·8	22·8	18·3	13·3	8·3	2·2	11·7°C
	(2825′)	R	33·0	30·5	33·0	33·0	48·3	25·4	12·7	10·2	17·8	23·0	30·5	48·3	345·5 mm
14 BAGHDAD	34 m	T	10·0	11·7	15·0	21·7	27·8	31·7	33·9	33·9	30·6	24·4	17·8	11·7	22·8°C
	(110′)	R	23·0	25 5	28·0	12·7	2·5	0·0	0·0	0·0	0·0	2·6	20·3	25·4	140 mm
15 CALCUTTA	6 m	T	19·4	22·2	27·2	30·0	30·0	29·4	28·9	28·3	28·9	27·8	23·3	19·4	26·1°C
	(20′)	R	10·2	30·5	35·6	43·2	139·7	297·2	325·0	327·7	251·5	114·3	20·3	5·0	1600·0 mm

STATION	Altitude	Temp.°C Rain mm	J	F	M	A	M	J	J	A	S	O	N	D	Average Total
16 COLOMBO	8 m	T	26·1	26·1	27·2	27·8	28·3	27·2	27·2	27·2	27·2	26·7	26·1	25·6	27·2°C
	(25′)	R	89·0	68·6	147·3	231·2	370·8	223·5	134·6	109·2	160·0	347·9	315·0	147·3	2344·4 mm
17 IRKUTSK	466 m	T	−21·1	−18·3	−10·6	−0·6	7·2	13·3	15·6	14·4	7·8	−0·6	−11·7	−15·6	−2·2°C
	(1530′)	R	12·7	10·2	7·6	15·2	33·0	56·0	78·6	71·1	43·2	17·8	15·2	15·2	378·5 mm
18 KATMANDU	1338 m	T	10·0	11·7	16·1	20·0	22·8	24·4	24·4	23·9	23·3	20·0	15·0	11·1	18·9°C
	(4390′)	R	15·2	40·6	22·9	58·4	122·0	246·3	373·4	345·4	154·9	38·1	7·6	2·5	1427·5 mm
19 SHANGHAI	7 m	T	3·9	4·4	8·3	14·4	20·0	23·3	27·8	27·8	23·3	18·3	12·2	6·7	16·1°C
	(23′)	R	48·3	58·4	83·8	93·9	94·0	180·3	147·3	142·2	129·5	71·1	50·8	35·6	1135·5 mm
20 SINGAPORE	10 m	T	26·1	26·7	27·2	27·3	27·7	27·2	27·2	27·2	27·2	26·7	26·7	26·7	26·7°C
	(33′)	R	251·5	172·7	193·0	187·9	172·7	172·7	170·2	195·6	177·8	208·3	254·0	256·5	2413·0 mm
21 TOKYO	6 m	T	3·3	3·9	7·2	12·2	16·7	20·6	24·4	26·1	22·2	16·7	10·6	5·6	14·4°C
	(20′)	R	48·3	73·7	106·7	134·6	147·3	166·0	142·2	152·4	233·7	208·3	96·5	55·9	1564·5 mm
22 URUMCHI	906 m	T	−16·1	−13·9	−6·1	8·9	15·0	18·9	21·1	20·0	14·4	4·4	−6·1	−11·1	4·4°C
	(2972′)	R	15·2	7·6	12·7	38·0	27·9	38·1	17·8	25·4	15·2	43·2	40·6	10·2	292·0 mm
23 VERKHOYANSK	100 m	T	−50·6	−44·4	−32·2	−15·0	0·0	12·2	13·3	9·4	1·7	−15·6	−37·2	−47·8	−17·2°C
	(328′)	R	5·0	5·1	2·5	5·1	7·6	22·9	27·9	25·4	12·7	7·6	7·6	5·1	134·5 mm
24 VLADIVOSTOK	29 m	T	−14·4	−10·0	−3·3	4·4	9·4	13·9	18·3	20·6	16·1	8·9	−1·1	−10·0	4·4°C
	(95′)	R	7·6	10·3	17·8	30·5	53·3	73·7	83·8	119·4	109·2	48·3	30·5	15·2	599·5 mm

AFRICA

STATION	Altitude	Temp.°C Rain mm	J	F	M	A	M	J	J	A	S	O	N	D	Average Total
25 ACCRA	27 m	T	26·7	27·8	27·8	27·8	27·2	26·1	25·0	23·9	25·0	26·1	27·2	27·8	26·7°C
	(88′)	R	15·3	33·0	55·9	81·3	142·2	177·8	45·7	15·2	35·6	63·5	35·6	22·9	724·0 mm
26 ALGIERS	59 m	T	12·2	12·8	13·9	16·7	18·9	21·7	24·4	25·6	23·9	20·0	16·1	12·8	18·3°C
	(194′)	R	111·8	83·8	73·7	40·6	45·7	15·2	0·0	5·1	40·6	78·7	129·5	137·2	762·0 mm
27 CAIRO	116 m	T	13·3	14·4	17·2	21·1	25·0	27·8	28·3	28·3	26·1	23·9	20·0	15·0	22·2°C
	(380′)	R	5·2	5·1	5·1	2·5	0·0	0·0	0·0	0·0	0·0	0·0	2·5	5·1	25·5 mm
28 CAPE TOWN	17 m	T	20·6	20·6	19·4	16·7	14·4	12·8	12·2	12·8	13·9	16·1	17·8	19·4	16·7°C
	(56′)	R	15·2	7·6	17·8	48·3	78·7	83·8	88·9	66·0	43·2	30·5	17·8	10·2	508·0 mm
29 DURBAN	5 m	T	23·9	23·9	23·3	21·7	18·9	17·8	16·7	17·2	18·9	20·0	21·7	22·8	20·6°C
	(16′)	R	109·3	121·9	129·5	76·2	50·8	33·0	27·9	38·1	71·1	109·2	121·9	119·4	1008·5 mm
30 KANO	464 m	T	21·1	23·9	27·8	31·1	30·6	28·3	26·1	25·0	26·1	26·7	24·4	22·2	26·1°C
	(1523′)	R	0·0	0·0	2·5	10·2	68·7	116·8	205·8	310·0	142·2	12·8	0·0	0·0	869·0 mm
31 KHARTOUM	390 m	T	23·3	25·0	28·3	31·1	33·3	33·3	31·7	30·6	31·7	31·7	28·3	22·2	29·4°C
	(1279′)	R	0·0	0·0	0·0	0·0	2·6	7·6	53·3	71·1	17·8	5·1	0·0	0·0	157·5 mm

STATION	Altitude	Temp.°C Rain mm	J	F	M	A	M	J	J	A	S	O	N	D	Average Total
32 KISANGANI	418 m	T	25·6	25·6	25·6	26·1	25·6	25·0	23·9	23·9	24·4	25·0	25·0	25·0	25·0°C
	(1370′)	R	53·4	83·8	177·8	157·5	137·2	114·3	132·1	165·1	183·0	218·4	198·1	83·8	1704·5 mm
33 NAIROBI	1820 m	T	18·3	19·4	19·4	18·9	17·8	16·7	15·6	16·1	17·2	18·3	18·3	18·3	18·3°C
	(5971′)	R	38·1	63·5	124·5	210·8	157·5	45·6	15·2	22·9	30·5	53·3	109·2	86·4	957·5 mm
34 PRETORIA	1369 m	T	21·1	21·1	19·4	16·7	13·3	11·1	11·1	13·3	17·2	19·4	20·0	21·1	17·2°C
	(4490′)	R	127·0	109·2	114·3	43·2	22·9	15·2	7·6	5·1	20·3	60·0	132·1	132·1	785·0 mm
35 SALISBURY	1474 m	T	20·6	20·6	20·0	18·9	16·7	13·9	13·9	15·6	18·9	21·1	21·1	20·6	18·3°C
	(4831′)	R	195·6	177·8	116·8	27·9	12·7	2·6	0·0	2·5	5·1	27·9	96·5	162·6	828·0 mm
36 WALVIS BAY	7 m	T	18·9	19·4	18·9	18·3	17·2	16·1	14·4	13·9	13·9	17·8	16·7	17·8	17·2°C
	(24′)	R	0·1	5·1	7·6	2·5	2·5	0·0	0·0	2·5	0·0	0·0	0·0	0·0	20·3 mm

NORTH AMERICA

STATION	Altitude	Temp.°C Rain mm	J	F	M	A	M	J	J	A	S	O	N	D	Average Total
37 AKLAVIK	9 m	T	–27·8	–26·7	–22·2	–12·8	–0·6	9·4	13·3	10·0	3·3	–6·7	–19·4	–26·7	–8·9°C
	(30′)	R	12·7	12·7	10·2	12·6	12·7	20·3	35·6	35·6	22·9	22·9	30·3	10·2	228·5 mm
38 CHICAGO	251 m	T	–3·9	–2·8	2·2	8·3	13·9	19·4	22·8	22·2	18·3	12·2	4·4	–1·1	10·0°C
	(823′)	R	50·6	50·8	66·0	71·1	86·4	88·9	83·8	81·3	78·7	66·0	60·9	50·8	835·5 mm
39 CHURCHILL	13 m	T	–28·3	–26·7	–21·1	–10·0	–1·1	6·1	11·7	11·1	5·0	–2·8	–14·4	–23·9	–8·3°C
	(43′)	R	12·8	15·2	22·9	22·9	22·9	48·3	55·9	68·6	58·4	35·6	25·4	17·8	406·5 mm
40 DALLAS	156 m	T	7·2	10·0	13·3	18·3	22·2	26·7	28·9	28·9	25·6	19·4	13·3	8·3	18·3°C
	(512′)	R	63·5	61·0	83·8	106·7	114·3	96·5	71·7	76·2	68·6	71·1	68·6	63·5	945·0 mm
41 HALIFAX	30 m	T	–5·0	–5·0	–1·1	3·9	10·0	14·4	17·8	18·3	14·4	9·4	3·9	–2·2	6·7°C
	(99′)	R	137·2	109·2	124·5	114·3	104·1	101·6	96·5	111·8	104·1	137·2	134·6	137·2	1412·2 mm
42 MEXICO CITY	2309 m	T	12·2	13·3	16·1	17·8	18·9	18·3	17·2	17·2	17·2	15·6	13·9	12·2	15·6°C
	(7575′)	R	12·8	5·2	10·2	20·3	53·3	119·4	170·2	152·4	129·5	50·8	17·8	7·6	747·0 mm
43 MIAMI	8 m	T	19·4	20·0	21·7	22·8	25·0	26·7	27·8	27·8	27·2	25·0	22·2	20·6	23·9°C
	(25′)	R	71·2	53·3	63·5	81·3	172·7	177·8	154·9	160·0	203·2	233·7	71·1	50·8	1516·5 mm
44 NEW YORK	96 m	T	–1·1	–0·6	2·8	10·0	15·6	20·0	23·3	22·8	20·6	15·0	6·7	1·1	8·3°C
	(314′)	R	93·9	96·5	91·4	81·3	81·3	83·8	106·7	109·2	86·4	88·9	76·2	91·4	1092·2 mm
45 OTTAWA	103 m	T	–11·1	–11·1	–4·4	5·0	12·8	18·3	20·6	18·9	14·4	7·2	0·0	–8·9	5·0°C
	(339′)	R	73·7	55·9	71·1	68·6	63·5	88·9	86·4	66·0	81·3	73·7	76·2	66·0	871·2 mm
46 PHOENIX	330 m	T	11·1	13·3	16·1	19·4	23·9	29·4	32·2	31·1	28·3	21·7	15·6	11·7	21·1°C
	(1083′)	R	20·3	20·3	17·8	10·2	2·5	2·5	25·4	25·4	17·8	10·2	15·2	22·9	190·5 mm

STATION	Altitude	Temp. °C Rain mm	J	F	M	A	M	J	J	A	S	O	N	D	Average Total
47 SACRAMENTO	21 m	T	7·2	10·0	12·8	14·4	17·8	21·1	23·3	22·8	21·1	16·7	12·2	7·8	15·6°C
	(69′)	R	96·5	71·1	71·1	38·1	20·3	2·5	0·0	0·0	7·6	20·3	48·3	96·5	472·3 mm
48 VANCOUVER	14 m	T	2·2	3·9	6·1	9·4	12·8	15·6	17·8	17·2	13·9	10·0	6·1	3·9	10·0°C
	(45′)	R	218·5	147·3	127·0	83·8	71·1	63·5	30·5	43·2	91·4	147·3	210·8	223·5	1458·0 mm

SOUTH AMERICA

STATION	Altitude	Temp. °C Rain mm	J	F	M	A	M	J	J	A	S	O	N	D	Average Total
49 ANTOFAGASTA	94 m	T	20·6	20·6	19·4	17·8	16·1	14·4	13·9	13·9	14·4	15·6	17·2	18·9	17·2°C
	(308′)	R	0·0	0·0	0·0	0·0	0·0	2·5	5·1	2·5	0·0	2·5	0·0	0·0	12·6 mm
50 BELEM	13 m	T	26·1	26·1	26·7	26·7	26·7	26·7	26·1	26·1	26·7	26·7	26·7	26·7	26·7°C
	(42′)	R	317·5	358·2	358·1	320·0	259·1	170·2	149·9	111·8	88·9	83·8	66·0	154·9	2438·5 mm
51 BUENOS AIRES	27 m	T	23·3	22·8	20·6	16·7	12·8	9·4	10·0	10·6	12·8	15·0	18·9	21·7	16·1°C
	(89′)	R	78·7	71·1	109·2	88·9	76·2	60·9	55·9	60·9	78·7	86·4	83·8	99·2	950·0 mm
52 LA PAZ	3658 m	T	11·7	11·7	11·7	11·1	10·0	8·9	8·3	9·4	10·6	11·7	12·2	11·7	10·6°C
	(12,001′)	R	114·3	106·7	66·0	33·0	12·7	7·6	10·2	12·7	27·9	40·6	48·3	93·9	574·0 mm
53 MANAOS	44 m	T	27·2	27·2	27·2	27·2	27·2	27·2	27·8	28·3	28·3	28·9	28·3	27·8	27·8°C
	(144′)	R	248·9	231·1	261·6	220·9	170·2	83·8	58·4	38·1	45·7	106·7	142·2	203·2	1811·0 mm
54 PORT OF SPAIN	20 m	T	25·6	25·6	25·6	26·1	26·7	26·7	26·1	26·1	26·7	26·7	26·7	25·6	26·1°C
	(67′)	R	68·6	40·6	45·7	53·3	93·9	193·0	218·4	246·4	193·0	170·2	182·9	124·5	1630·7 mm
55 PORT STANLEY	2 m	T	9·4	8·9	7·8	6·1	3·9	2·2	1·7	2·2	3·9	5·0	6·7	7·8	5·6°C
	(6′)	R	71·1	58·4	63·5	66·0	66·0	53·3	50·8	50·8	38·1	40·6	50·8	71·1	680·7 mm
56 RIO DE JANEIRO	61 m	T	25·6	26·1	25·0	23·3	21·7	21·1	20·6	21·1	21·1	21·7	22·8	24·4	22·8°C
	(201′)	R	124·5	121·9	129·5	106·7	78·7	53·3	40·6	43·2	66·0	78·7	104·1	137·2	1082·0 mm
57 VALPARAISO	41 m	T	17·8	17·8	16·7	15·0	13·3	12·2	11·7	12·2	12·8	13·9	15·6	16·7	14·4°C
	(135′)	R	2·5	0·0	7·6	15·2	104·1	149·8	99·1	73·7	33·0	10·2	5·1	5·1	505·4 mm

AUSTRALASIA

STATION	Altitude	Temp. °C Rain mm	J	F	M	A	M	J	J	A	S	O	N	D	Average Total
58 ADELAIDE	43 m	T	22·8	23·3	21·1	17·8	14·4	12·2	11·1	12·2	13·9	16·7	19·4	21·7	17·2°C
	(140′)	R	20·4	17·8	25·4	45·7	68·6	76·2	66·0	66·0	53·3	43·2	27·9	25·4	536·0 mm
59 ALICE SPRINGS	579 m	T	28·3	27·8	24·4	19·4	15·0	12·2	11·7	14·4	18·3	22·8	25·0	27·8	20·6°C
	(1900′)	R	43·2	33·0	27·9	10·2	15·2	12·7	7·6	7·6	7·6	17·8	30·5	38·1	251·5 mm

STATION	Altitude	Temp. °C Rain mm	J	F	M	A	M	J	J	A	S	O	N	D	Average Total
60 AUCKLAND	26 m	T	18·9	18·9	18·3	16·1	13·3	11·7	10·6	11·1	12·2	13·9	15·6	17·2	15·0°C
	(85′)	R	78·7	93·8	81·3	96·5	127·0	137·2	144·8	116·8	101·6	101·6	88·9	78·7	1247·0 mm
61 BRISBANE	42 m	T	25·0	24·4	23·3	21·1	18·3	15·6	14·4	15·6	18·3	21·1	22·8	24·4	20·6°C
	(137′)	R	162·6	160·0	144·8	94·0	71·1	66·0	55·9	48·3	48·3	63·5	93·9	127·0	1135·5 mm
62 DARWIN	30 m	T	28·3	28·3	28·9	28·9	27·8	25·6	25·0	26·1	27·8	29·4	30·0	29·4	27·8°C
	(97′)	R	386·2	312·4	254·0	96·5	15·2	2·5	0·0	2·5	12·7	50·8	119·4	238·8	1491·0 mm
63 DUNEDIN	73 m	T	14·4	14·4	12·8	11·1	8·3	6·7	5·6	6·7	8·9	10·0	11·7	13·3	10·6°C
	(240′)	R	86·4	71·1	76·2	71·1	81·3	81·3	78·7	76·2	68·6	76·2	81·3	88·9	937·3 mm
64 HOBART	54 m	T	16·7	16·7	15·0	12·8	10·6	8·3	7·8	8·9	10·6	12·2	13·9	15·6	12·2°C
	(177′)	R	48·3	38·1	45·7	48·3	45·7	55·8	53·3	48·3	53·3	58·4	60·9	53·3	609·5 mm
65 PERTH	60 m	T	23·3	23·3	21·7	18·9	16·1	13·9	12·8	13·3	14·4	16·1	18·9	21·7	17·8°C
	(197′)	R	7·6	10·2	20·3	43·2	129·5	180·3	170·2	144·8	86·4	56·0	20·3	12·7	881·5 mm
66 SYDNEY	42 m	T	21·7	21·7	20·6	17·8	15·0	12·2	11·7	12·8	15·0	17·2	19·4	21·1	17·2°C
	(138′)	R	88·8	101·6	127·0	134·6	127·0	116·8	116·8	76·2	73·7	71·1	73·7	73·7	1181·0 mm

III. Examination Questions

The following selection of examination questions will provide useful practice in using some of the information covered in this book, but of course they are not a substitute for past papers for the particular examination which you propose to sit.

Although most of the questions deal only with climate, every examination contains questions in which climatic information has to be used along with various other aspects of human and physical geography, and a few questions of this type have been included.

Questions 1–16 are at Advanced Level.
Questions 17–25 are at the Higher Grade.
Questions 26–42 are at the Ordinary Level.

Advanced Level

From Advanced Level.
Joint Matriculation Board examination papers.

1. Describe the character and distribution of *three* major types of 'natural' vegetation in any one continent. Select *one* of these types and show how the character of its vegetation depends on the various elements of the environment.

2. Describe and explain the distribution of mean sea-level temperatures in January and July for any *one* continent. Illustrate how the distribution pattern is modified by the influence of relief.

3. Discuss the relationship between man and his environment in areas that are predominantly given over to *either* nomadic herding *or* subsistence farming. Illustrate your discussion by reference to *two* contrasted areas or societies.

4. Give a reasoned account of the variations in mean annual rainfall and its seasonal distribution in the U.S.A. and Canada.

5. Discuss the major contrasts in economic development between the areas of tropical and temperate grassland in Latin America.

6. Give a reasoned account of the mean annual rainfall and its seasonal distribution in the Indian subcontinent.

7. Study the sea route shown on the map provided and suggest, with reasons, the possible temperature, rainfall and wind conditions for *four* of the named positions on the route at the dates indicated.

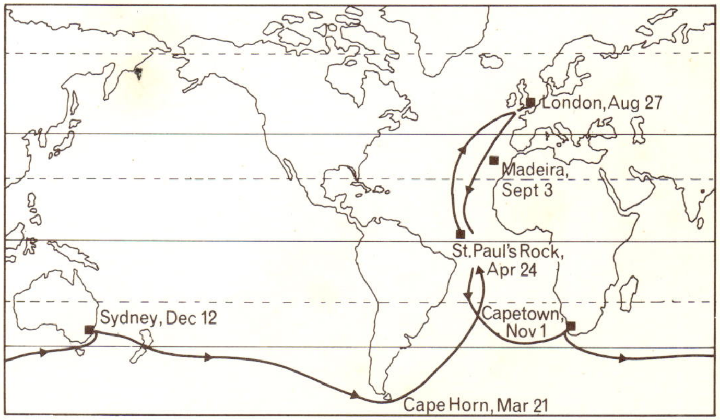

I

8. Discuss the effects of the major types of air mass on the climate of the British Isles.

9. How has climate influenced the agricultural development of the West Indian islands?

10. Describe the climatic transition *either* (*a*) from the West African coastlands to the Sahara *or* (*b*) from the coast of Natal to that of South West Africa. Briefly outline the influence of climate on contrasts in natural vegetation and agriculture within the area.

From Oxford and Cambridge Schools Examination Board examination papers.

11. What changes might take place in the climates of the lands adjacent to the North Atlantic if the surface oceanic circulation were to be reversed?

12. Consider the part played by climate in creating (*a*) opportunities, and (*b*) problems, for agriculture in tropical South America.

13. Comment on the shortcomings of using the average (the arithmetic mean) in generalizing climatic data and suggest useful refinements and alternatives.

14. Under what conditions does fog occur?

15. What climatic data does a geographer need in the study of agricultural potential? Justify your selection.

From Southern Universities Joint Board examination paper.

16. With reference to the weather map:

(*a*) explain the position of the sea-water isotherm (10°C) off both the Atlantic and North Sea coasts;

(*b*) using correct Meteorological Office symbols, insert for each of the stations A, B, C and D the conditions of wind direction, wind speed (knots), cloud cover (eighths), temperature (°C) and weather that you think are likely to be experienced in view of the general pressure situation shown;

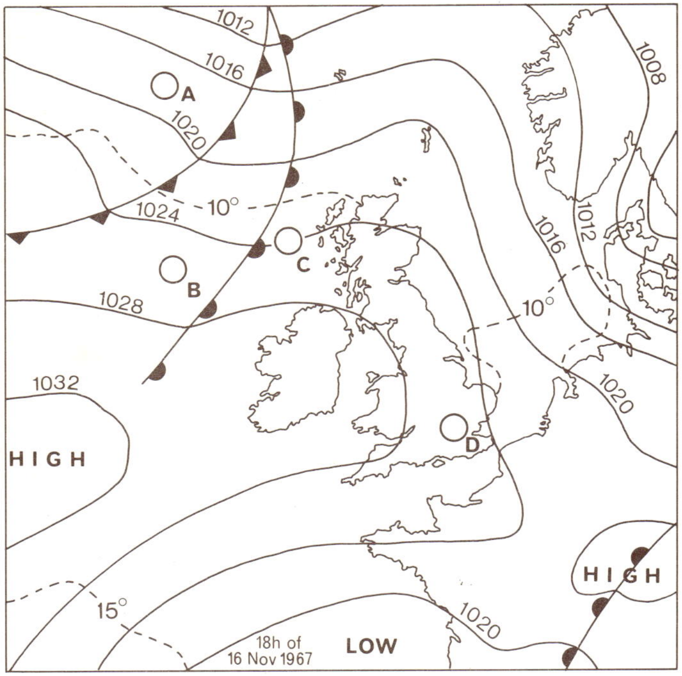

--10°-- sea water isotherm

(*c*) give full reasons for the weather conditions that you have shown at each of the stations A, B, C and D;

(*d*) attempt a reasoned forecast of weather conditions over the British Isles during the next 48 hours. (State clearly your assumptions of the direction of movement of the pressure systems.)

Higher Grade

From Scottish Certificate of Education Examination Board. Higher Grade papers.

17. The map of the world supplied marks the route taken by Sir Francis Chichester on his historic single-handed voyage round the world.

(*a*) Write an explanatory account of the route followed by Sir Francis in the light of global patterns of winds and currents. Annotate the map accordingly and give particular attention in your account to the Atlantic Ocean section of his journey.

(*b*) With reference to the North Atlantic Ocean explain the pattern of ocean currents and the fact that some are warm and others cold.

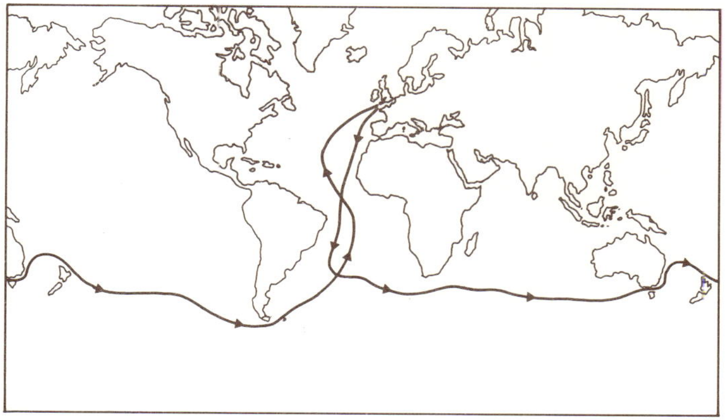

18. The following statistics relate to Bergen (Norway).

		J	F	M	A	M	J	J	A	S	O	N	D	Year
Mean Monthly Temperature	°F	34	34	35	42	49	55	58	58	53	45	39	35	44·6
	°C	1	1	2	6	9	13	14	14	12	7	4	2	7
Mean Monthly Rainfall	in	6·9	5·6	4·3	3·8	4·1	4·4	5·9	6·9	8·3	8·8	6·7	7·4	73·1
	mm	175	142	109	97	104	112	150	175	211	224	170	188	1857

(Altitude of weather station—66′.)

(*a*) Describe the factors which have had an effect on these temperature and rainfall figures.

(*b*) In what sense do these figures present an incomplete picture of the climate of Bergen?

19. Using *one* of the following continents to illustrate your answer, justify the statement that *most* rainfall regimes show a summer maximum, explaining supporting cases, and any exceptions: North America, *or* Asia, *or* Australia.

20. (*a*) The map of Scotland gives the mean annual rainfall for 40 meteorological stations. On the map the main continuous isohyets of 30″ and 60″ are partially drawn. Using the data given, complete these isohyets.
Explain, with specific examples, why the 60″ isohyet as drawn gives an incomplete picture of the distribution of rainfall of over 60″ in Scotland.

(*b*) How is the mean annual rainfall of any specific meteorological station determined?

(*c*) Using any recognized cartographic method, show the amount and seasonal distribution

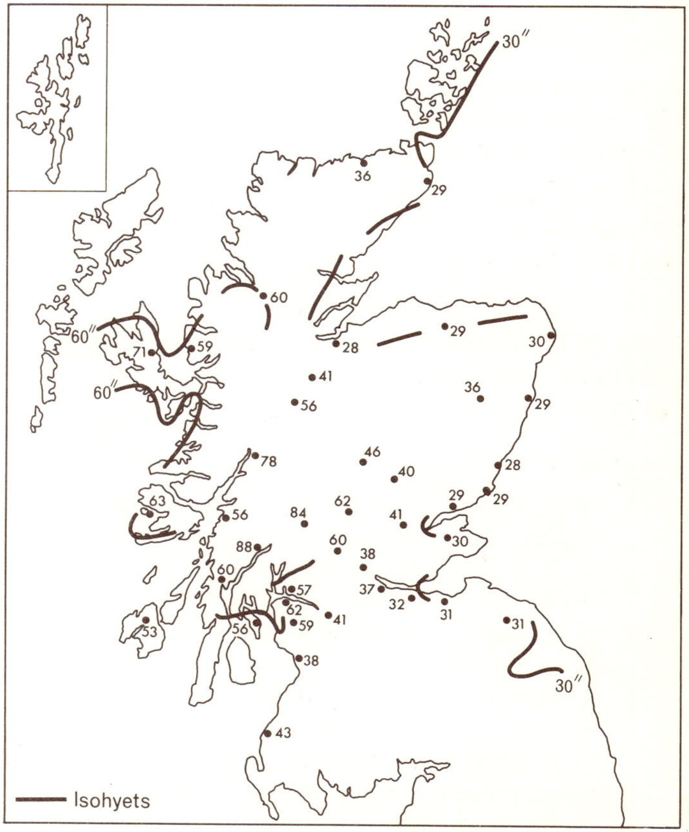

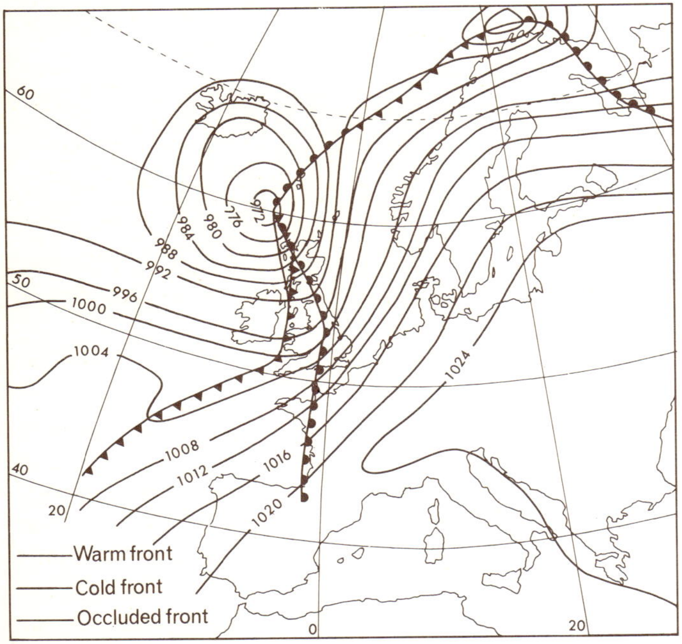

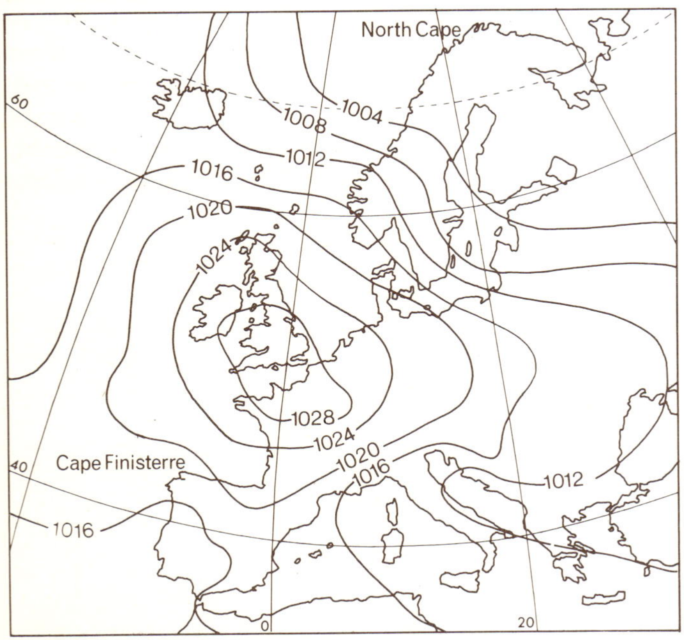

of rainfall at some indicated place within your Home Area. (Answer on squared paper.)

21. Answer *either* (*a*) *or* (*b*).

(*a*) After studying the synoptic charts provided, describe the general weather conditions associated with each system shown.

or

(*b*) Using the information given, indicate where, in your opinion, the following climatic stations are situated:

Station	Altitude	Temperature and Rainfall	J	F	M	A	M	J	J	A	S	O	N	D	Year
A	197′	Temp. °F	74	74	71	66	61	57	55	56	58	61	66	71	64
		Rainfall in	0·3	0·4	0·8	1·7	5·1	7·1	6·7	5·7	3·4	2·2	0·8	0·5	34·7
B	124′	Temp. °F	28	29	36	46	57	67	72	70	63	54	42	32	50
		Rainfall in	3·6	3·3	3·8	3·5	3·1	3·2	3·3	3·6	3·2	3·3	3·6	3·4	40·8
C	72′	Temp. °F	72	71	68	61	56	51	50	51	54	58	64	69	61
		Rainfall in	2·9	2·6	3·9	3·9	3·3	3·2	2·9	3·1	3·0	2·6	2·9	3·1	37·4
D	482′	Temp. °F	36	40	47	55	65	71	75	75	67	56	45	39	56
		Rainfall in	2·2	2·2	3·5	3·5	3·0	3·0	2·6	2·6	2·8	4·5	4·0	3·3	37·2

22. Attempt a reasoned classification of the climates that occur within the tropics.

23. Describe the major changes of the natural vegetation you would observe during a train journey across Canada and relate them to changes in the climate.

24. Write an explanatory account of the climatic regime of tropical grasslands and describe the changes throughout the year in the appearance of these areas. Discuss, with examples, the effects of the climate on the farming of such regions.

25. The following climatic statistics refer to three places—X, Y, and Z. The altitude of X is 72 feet, of Y 138 feet, and of Z 4000 feet. Indicate the types of climate and in which parts of the world you would place them. State your reasons.

		Jan	Feb	Mar	Apr	May	June	July	Aug	Sep	Oct	Nov	Dec	Year
X	°F	34	34	36	42	49	55	58	57	52	45	39	36	45
	in	9·0	6·6	6·2	4·3	4·7	4·1	5·7	7·8	9·2	9·3	8·5	8·9	84·3
Y	°F	72	71	69	65	59	55	53	55	59	64	67	70	63
	in	3·7	4·3	4·8	5·6	5·1	4·8	4·8	3·0	2·9	3·2	2·9	2·9	47·9
Z	°F	34	42	48	61	71	80	85	83	77	66	51	42	62
	in	1·6	1·0	1·9	1·4	0·5	0·1	0·2	0	0·1	0·3	1·0	1·3	9·3

Ordinary Level

From Joint Matriculation Board examination papers.

26. (*a*) On the map, (i) shade and label 'H' a land area which in winter experiences high barometric pressure; (ii) draw a line to indicate a path commonly followed by depressions in temperate latitudes and label it 'D'; (iii) insert and name the Tropic of Cancer and the Tropic of Capricorn; (iv) draw an arrow to indicate the direction of a cold coastal current in the North Pacific Ocean.

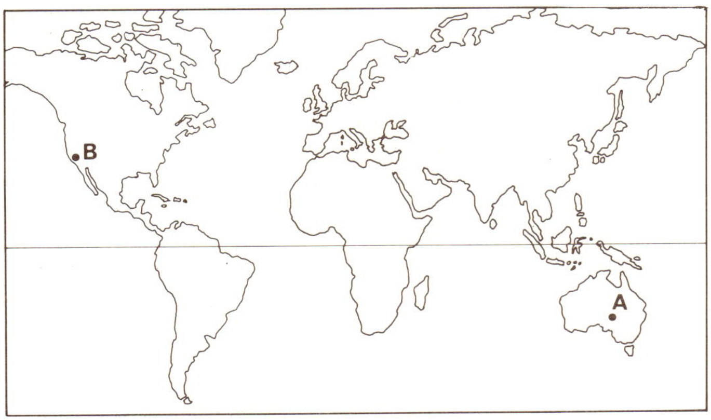

(*b*) A and B are weather stations marked on the map. The following climatic data refer to stations A and B.

		Station	Height in feet	J	F	M	A	M	J	J	A	S	O	N	D	
Mean Monthly Temperatures	°C	A	250	28	28	25	20	15	12	11	13	17	21	25	28	Range 17
	°F			83	83	76	67	59	54	52	56	62	70	77	81	Range 31
	°C	B	71	8	10	12	14	17	21	23	22	21	17	12	8	Range 15
	°F			46	50	54	58	63	69	73	72	69	62	53	46	Range 27
Rainfall	inches	A		0·5	0·4	0·8	0·4	0·4	0·7	0·3	0·3	0·4	0·3	0·4	0·3	Total 5·2
		B		3·8	2·9	3·0	1·6	0·8	0·1	0	0	0·2	0·9	2·1	4·0	Total 19·4

Give reasons why (1) the rainfall is higher at B than at A; (2) the range of mean monthly temperatures is greater at A than at B.

(*c*) Why is the weather in the British Isles so changeable?

27. In each of the following areas the inhabitants live under difficult conditions: Arabia, the northern forests of Canada, Northern Chile, the Siberian tundra, the Tibetan plateau. Choose *three* of these areas and for each (i) draw a sketch-map to show its location; (ii) explain why the conditions of life are difficult.

28. (*a*) On the map, (i) shade and label 'D' an extensive area of desert, and 'F' an area (outside the British Isles) subject to widespread and regular river floods; (ii) print 'A' over an area where avalanches are experienced, 'H' over an area where hurricanes occur, 'I' over an area where icebergs are a hazard, and 'S' over an area where soil erosion is a problem.

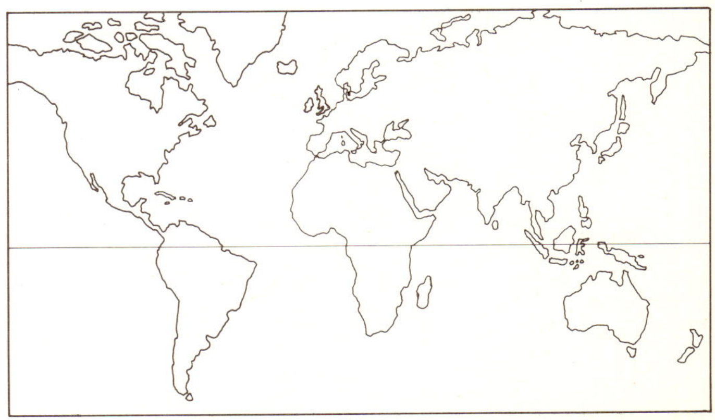

(*b*) Give reasons for (i) the desert conditions in the area you have marked 'D', (ii) the river floods of the area you have marked 'F'.

(*c*) The following climatic data refer to a sea-level station.

		Mean Monthly Temperatures												
Months		J	F	M	A	M	J	J	A	S	O	N	D	
	°C	27	27	27	27	27	27	27	28	28	28	28	27	Range 1
	°F	80	80	80	80	80	80	81	82	83	83	82	81	Range 3
Rainfall (in)		9·2	9·0	9·6	8·5	7·0	3·6	2·2	1·4	2·0	4·1	5·5	7·7	Total 69·8

(i) On the map shade and label 'X' *one* area where such conditions could occur.

(ii) For the area you have shaded show how such conditions have affected economic development.

29. Select *two* of the following types of vegetation: coniferous forest, equatorial forest, tropical grassland, tundra. For *each* (i) draw a sketch-map to show the location of a large area where it occurs; (ii) describe the climatic conditions of this area; (iii) show how the vegetation and climate have affected the economic development of the area.

From Oxford and Cambridge examination papers.

30. The following graphs of temperature and rainfall refer to three climatic stations *P*, *Q* and *R*.

(*a*) Describe the climate of each station as indicated by the graphs.

(*b*) Name the type of climate at each station.

(*c*) Choose *one* of the three types of climate represented and (i) locate by means of sketch-maps *two*

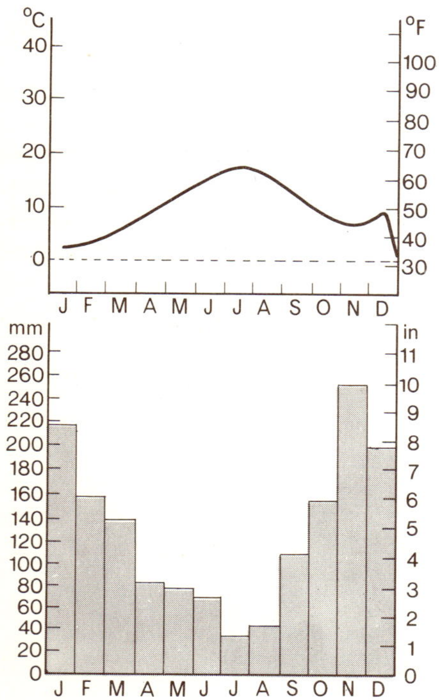

P

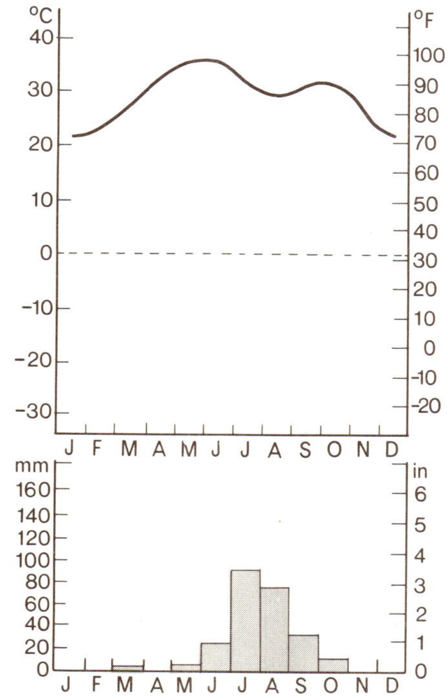

Q

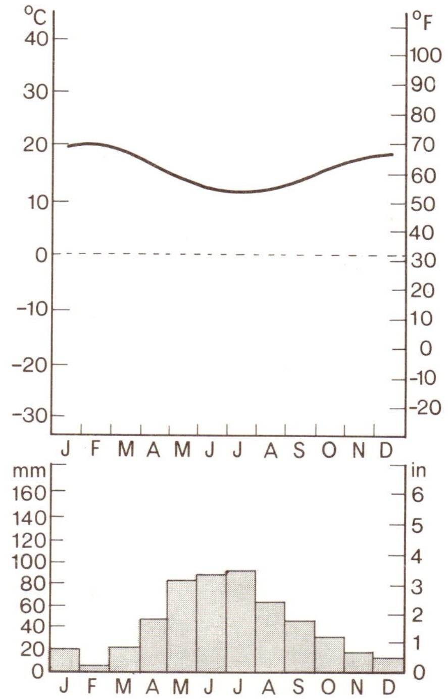

R

areas where this type of climate is found; (ii) describe the natural vegetation of *one* of these areas.

31. Expand and explain *two* of the following statements:

(*a*) Much of the interior of western Europe has moderate to light summer rainfall.

(*b*) In western Europe the pattern of the sea-level isotherms for January contrasts strikingly with that for July.

(*c*) The north-west of France has a climate of equable temperatures and moderate rainfall at all seasons.

32. Describe how air temperature, humidity and winds are recorded at a school weather station. Explain how this information may be mapped and used in the geography room.

33. Describe and account for the main features of the climate of *two* of the following: (*a*) California; (*b*) Ceylon; (*c*) the Congo Basin; (*d*) Greenland; (*e*) eastern Siberia.

34. The tables below give, for three lowland stations in England, the average monthly temperature and rainfall, together with their maximum and minimum recorded temperatures.

Table 1 gives the temperatures in degrees Fahrenheit and the rainfall in inches. Table 2 gives the

Table **1**

	J	F	M	A	M	J	Jy	A	S	O	N	D			
Station *A*															
	39·6	39·7	42·0	45·6	51·5	56·1	59·2	58·5	54·6	49·2	43·4	41·0	(°F)	Max.	91
														Min.	0
	6·4	4·5	3·6	3·3	3·1	3·4	4·3	5·2	4·9	6·4	6·1	6·7	(in)	Total	58
Station *B*															
	38·8	39·5	42·6	47·0	53·6	58·5	62·1	61·7	57·2	50·1	43·3	39·8	(°F)	Max.	96
														Min.	1
	2·0	1·3	1·4	1·7	1·7	1·6	2·2	2·0	1·9	2·0	2·0	1·9	(in)	Total	22
Station *C*															
	45·1	45·1	46·7	49·7	54·3	58·2	61·4	62·1	59·4	54·2	49·3	46·3	(°F)	Max.	85
														Min.	17
	5·0	3·5	3·3	2·5	2·4	2·1	2·7	3·0	2·9	4·1	4·8	5·2	(in)	Total	42

Table **2**

	J	F	M	A	M	J	Jy	A	S	O	N	D			
Station *A*															
	4·2	4·3	5·5	7·6	10·8	13·5	15·2	14·7	12·6	9·6	6·3	5·0	(°C)	Max.	33
														Min.	−18
	162	114	91	84	79	86	109	132	124	162	155	170	(mm)	Total	1473
Station *B*															
	3·7	4·1	5·7	8·3	12·0	14·7	16·8	16·6	14·1	10·0	6·3	4·3	(°C)	Max.	36
														Min.	−17·2
	51	33	36	43	43	41	56	51	48	51	51	48	(mm)	Total	559
Station *C*															
	7·3	7·3	8·2	9·8	12·4	14·6	16·3	16·8	15·3	12·3	9·7	7·9	(°C)	Max.	29
														Min.	−8·3
	127	89	84	63	61	53	69	76	74	104	122	132	(mm)	Total	1069

temperatures in degrees Centigrade and the rainfall in millimetres. Candidates may use either set of figures.

(*a*) Station *A* is the most northerly of the three stations. Station *C* is on the coast.

(i) How do the figures support these two statements?

(ii) Giving your reasons, suggest a location for *A* and for *C*.

(*b*) The rainfall total of station *B* is the lowest of the three; the monthly rainfall figures for *B* also show one marked difference in pattern from those of the other stations. What is that difference? How do you explain it?

35. (*a*) What is the Föhn wind, and why is it important in valleys in Switzerland?

(*b*) Pictures of the lower Rhône valley in France often show lines of cypresses across the valley. Obviously these trees have been planted. Why?

(*c*) At Cologne, roughly in the same latitude as London, the Rhine is normally icebound for about twenty days each winter. How do you explain this?

From Southern Universities examination papers.

36. Explain the nature of cyclonic depressions and describe the weather conditions associated with them in Britain.

37. Locate by sketch map one of the major equatorial rain forests of the world and:
(*a*) describe the characteristic natural vegetation;
(*b*) explain the patterns of temperature and rainfall;
(*c*) say how vegetation and climate have influenced economic development.

38. (*a*) In what ways do the distributions of pressure and temperature shown on the two synoptic charts differ?
(*b*) How does a knowledge of air masses help to explain some of the weather conditions shown on the charts? Illustrate your answer by reference to specific stations.
(*c*) What changes are likely to occur in the weather of the British Isles given an easterly movement of the pressure systems shown on each of the charts?

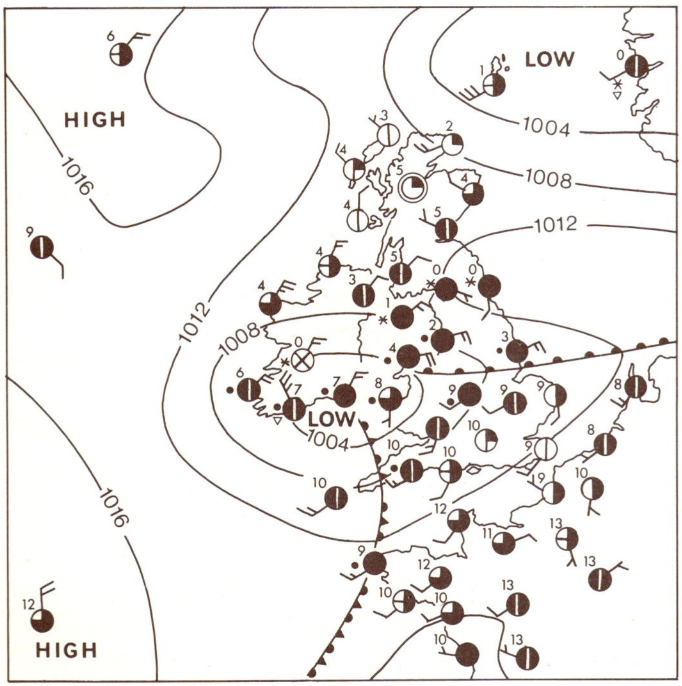

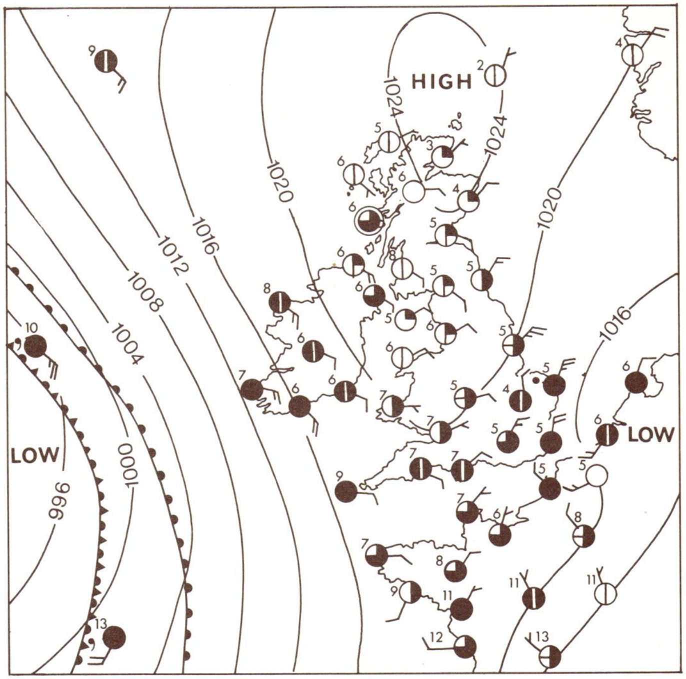

From Scottish Certificate of Education examination papers.

39. (*a*) Give *two* basic differences between taiga and selva forests.
(*b*) Describe briefly the climatic conditions (i.e. seasonal temperatures and precipitation) associated with each forest area.
(*c*) Write a short account of world trade in timber and the reasons for the increase in the use of this basic raw material.

40. Refer to the map of the world provided.
A cargo ship follows the route shown by the line, calling at ports numbered 1 to 7 *in that order*.
(*a*) Name the ports 1 to 7.
(*b*) From the following list choose the *two* most likely items of cargo to be loaded at each of the ports 1 to 6, and print their initials after the name of the port:

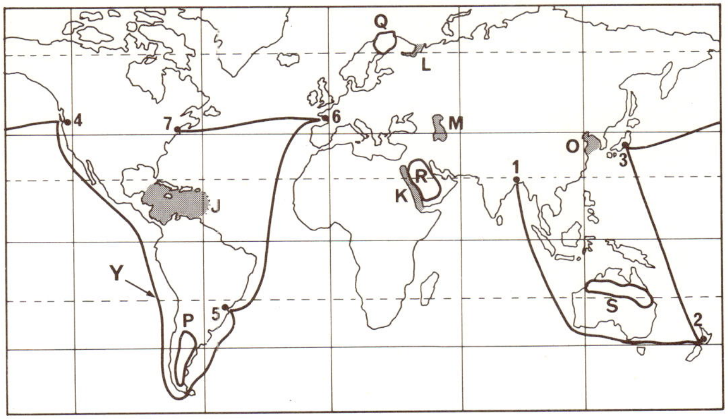

wool and butter (WB); tea and jute (TJ); coffee and cotton (CC); timber and tinned salmon (TS); cars and wines (CW); silks and cameras (SC).

(*c*) Name *two* ocean currents encountered between ports 5 and 6.

(*d*) Name the winds which predominate for most of the year at ports 2 and 4.

(*e*) The ship crosses the International Date Line between ports 3 and 4: enter in the given space either **DR** for 'date repeated' or **DM** for 'date missed out' as is appropriate for the ship's journey.

(*f*) Write down the date when the noonday sun is observed to be overhead at point Y.

(*g*) Name the five seas shaded and lettered J, K, L, M, O. Beside the appropriate two write **'hurricane'** and **'dhow'** where they might be encountered.

(*h*) Name the most important domesticated animal reared in each of the four outlined areas P, Q, R, S.

41. (*a*) From the map of Europe, name: island groups A and B; rivers C and D; countries E and F; canals G and H; island chain J and types of inlets along coasts K.

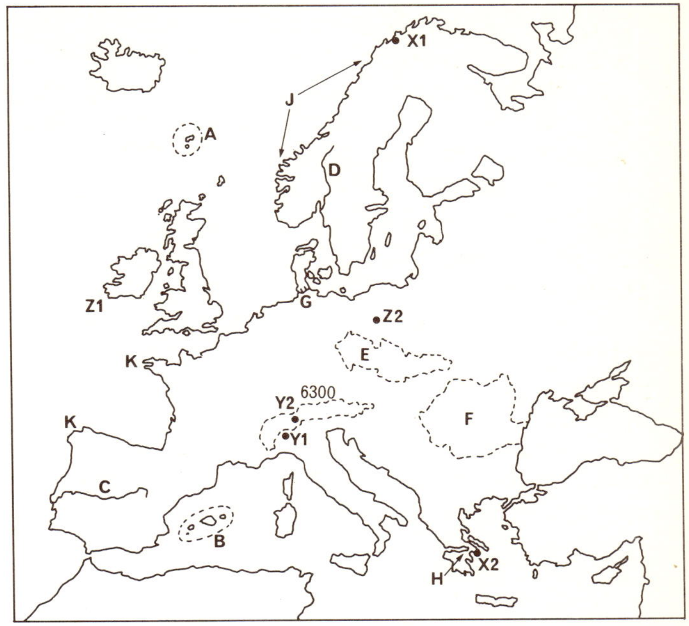

(*b*) Three pairs of climate stations—X1 and 2, Y1 and 2 and Z1 and 2—are shown on the map. Their mean July temperatures are as follows:

X1—52°F (11·1°C) X2—81°F (27·2°C)
Y1—75°F (23·9°C) Y2—54°F (12·2°C)
Z1—59°F (15°C) Z2—67°F (19·4°C)

Account briefly for the difference in temperature between each pair.

(*c*) Describe the farming activities of *one* of the following:

(i) eastern Jutland and the islands of Denmark;

(ii) the Rhine Rift Valley and adjacent mountain slopes;

(iii) the Plain of Aquitaine.

42. In order to produce a record of (*a*) mean monthly temperatures, and (*b*) seasonal distribution of rainfall, for a particular district, describe:

(i) the instruments required, and how they are set up;
(ii) what observations would be necessary, how they are made, and at what intervals;
(iii) what calculations would require to be made from the above readings; and
(iv) in what form, apart from tables, the records could be expressed.

IV. Conversion Graphs

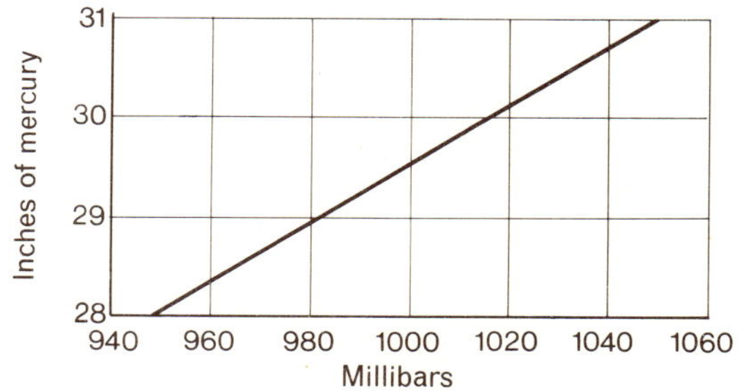

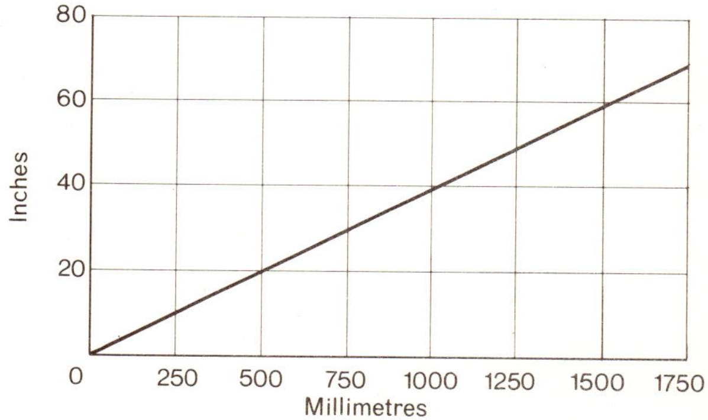

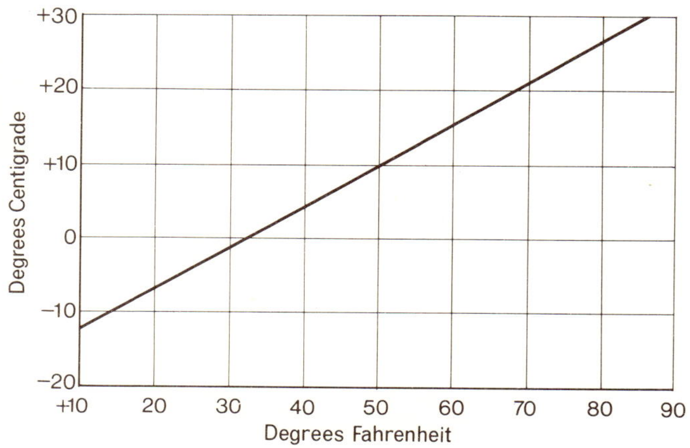

V. Some Reference Books

1. *Running a School Weather Station* by D. E. Pedgley (Royal Meteorological Society) (out of print)
2. Meteorological Office Statistics
   —Averages of Temperature for Great Britain and Northern Ireland.
   —Averages of Rainfall for Great Britain and Northern Ireland.
   —Tables of Temperature, Relative Humidity and Precipitation for the World.
   Part I —North America
   Part II —Central and South America
   Part III—Europe
   Part IV—Africa
   Part V —Asia
   Part VI—Australasia (not all in print)
3. *Meteorology and Climatology for Sixth Forms* by Ernest S. Gates (Harrap)
4. *Climatology* by A. Austin Miller (Methuen)
5. *Physical Geography and Climatology* by N. K. Horrocks (Longman)
6. *The Climates of the Continents* by W. G. Kendrew (Oxford University Press, 1922) (out of print)
7. *The Weather and Climate of the British Isles* by R. Kay Gresswell (Hulton Educational Publications)
8. *Climate and the British Scene* by G. Manley (Collins)
9. *World Vegetation* by D. Riley and A. Young (Cambridge University Press)
10. *The Climate of Glasgow* by J. A. Plant (H.M.S.O.) (out of print)
11. *Climate and Man* by L. H. Elliott (McGraw-Hill)
12. *An Introduction to Advanced Geography* by Briault and Hubbard (Longman)
13. *Climate, Soils, and Vegetation* by D. C. Money (University Tutorial Press)